Build it with Bales

A Step-by-Step Guide to Straw-Bale Construction

Version 1.0
Third Printing, October 1995

S. O. MacDonald

InHabitation Services
P.O. Box 58
Gila, New Mexico 88038

and

Matts Myhrman

Out On Bale
1037 E. Linden Street
Tucson, Arizona 85719

For all the women and men who, without prior knowledge of the technique, were inspired to create human shelter with bales. The rest of us have, at best, only improved a little on their idea.

> *One thing we do know, that we dare not forget, is that better solutions than ours have at times been made by people with much less information than we have.*
> **... Wendell Berry**

Library of Congress Cataloging-in-Publication Data

MacDonald, S. O.; Myhrman, M. A.
Build it with bales: a step-by-step guide to straw-bale construction

ISBN 0-9642821-0-0 (paperback)

Version 1.0, third printing, October 1995, 3000 copies

Cover and text printed, respectively, with soy-based ink on 50% and 100% recycled paper by West•Press, Tucson, AZ

Acknowledgements

We are grateful, above all, to our wives, Judy Knox and Nena I. MacDonald. When we procrastinated, they encouraged us gently to continue the work; when we digressed, they facilitated the necessary course corrections; when we despaired of having enough time to finish the task, they reminded us that we had all the time there was and helped us rearrange our priorities. They were, in fact, our valued partners.

Our thanks go also to Orien MacDonald, whose steady pen and three-dimensional vision transformed wavering lines, sometimes scrawled on beer-stained napkins, into the visual viscera of this collaboration.

Finally, we salute all the trail-breakers and compatriots, both named (Carol Ahlgren, David Bainbridge, Patti Brown, Chuck and Mary Bruner, Virginia Carabelli, Jake and Lucille Cross, Joanne DeHavillan, Lance Durand, David Eisenberg, Carol Escott, Hesh Fisk, Pliny Fisk, III, Barry Freeman, Jon Hammond, Steve Kemble, Bob Lanning, Jill Lorenzini, Gerry Matlock, Joe McCabe, Tony Perry, Jon Ruez, Bill and Athena Steen, Gary Strang, Brad Tatham, Ralph Towl, John Valentine, Catherine Wanek, and Roger Welsch) and unnamed (the list is even longer) who in various and nefarious ways informed, challenged, supported and inspired us (and the bale-building revival) as we traveled the Yellow-straw Road to Nebraska and beyond.

Disclaimer

The authors, Out On Bale, InHabitation Services, and the various individuals who have contributed details to this book make no statement, warranty, claim or representation, either expressed or implied, with respect to the construction details or methods described herein. Neither will they assume any liability for damages, losses or injuries that may arise from use of this publication. The details presented here are based on the best information available, but recent experience with the technique is finite and the information will not be appropriate for all conditions and/or climatic regimes. When in doubt, you are advised to consult with experienced people familiar with your local conditions and building codes.

Photo by Judy Knox

Matts' Three Bits

1. Judy Knox has been my valued partner and collaborator throughout our mutual involvement with the straw-bale revival. Whatever contributions we've made individually reflect that partnership.

2. A major blessing of this endeavor has been the opportunity to link minds with, bend elbows with, lock horns with, jaw-bone with, reason with, be persuaded by, learn about and learn from the poor-man's Marquis de Straw, my co-author Steve MacDonald. He does credit to his "grizzly" Alaskan - Finnish - Norwegian - Scottish heritage.

3. This document is truly a work-in-progress, as will be evident to anyone who puts its principles to use and discovers new and better ways to do things than are described here. **We entreat and encourage you** to help us make the next iteration of this opus more complete, more understandable and more useful. You will, thereby, become official members of the straw-bale "ghost writers in the sky" auxiliary. Send your suggestions to us via BIWB, 1037 E. Linden St., Tucson, AZ 85719.

...Matts Myhrman, Tucson

SOM's Seven Bits

1. *Keep it small.* How much space do you really need? Be honest. Be creative with your space. Pretend you're building storage on a ship. Small is easy to heat, and cool. It's easy to keep clean. It takes up fewer of the Earth's resources and takes up less of its space. You finish the job, at a lower cost, so you can devote yourself to more useful work. If your teenagers need distance, have them build their own outbuilding or addition. They need to learn the skills, anyway.

2. *Keep it simple.* Control your impulses to make your house a complicated "artsy" statement. Simple, small and rectangular houses are beautiful when made of straw and other natural materials. Let form follow function. Let go of the idea of having a perfectly square, flat and sharp-edged building. Again, spend the time and money you saved, by building with straw, in other ways – restore the river, help a neighbor, play with the kids.

3. *Build it yourself.* Trust yourself. You can do it, especially if you build with straw. And especially if you follow rules 1 and 2. Read building books and magazines. Ask your builder friends questions. Build it on paper and as a model first. Track the details. Use your common sense. Be creative with your mistakes. Don't be intimidated by the "experts". Get all the stuff together and host a straw-bale "barn raising".

4. *Stay out of debt.* Pay as you go. Assemble the parts as you have the money...and time. Make your barn raising a "potluck."

5. *Use local materials.* Use more rock and adobe, less concrete. Use locally-milled lumber and poles. Your neighbor needs the work and you need to know firsthand what demands you're asking of the forests and fields.

6. *Be energy conscious.* Build to maximize passive heating/cooling strategies. Superinsulate your ceiling. Stay off the electric power grid if you can. Put up a windmill. Use a solar pump. Build a composting toilet. Raise a garden. Throw out the television.

7. *Make yourself a home.* Don't just build yourself a house, make yourself a home. Stay where you are, if you can. Learn to be at home. Do no harm.

...Stephen O. MacDonald, Gila

Contents

Introduction

Why We Wrote This Guide

Somewhere in the yeasty revival of an "alternative" building method, the initial rapid pace of growth and change begins to slow down a little. Experimentation and learning will continue, but there now exists a body of knowledge that has already been validated by experience. Desktop publishing provides an economic way to start sharing such knowledge in printed form. It also allows future revision and expansion on a timely basis. So until someone chooses to publish the official, hard-backed, "complete and unabridged" bible of straw-bale construction, here's grist for the mill from two narrow-eyed fanatics.

How This Guide is Organized

We've divided the main body of the guide into two parts: Part One deals with the things you may want to do before you build; Part Two tracks through the building process. For the *Loadbearing Option* and the *Non-Loadbearing Option*, we focus on the "model" structure depicted in the overview drawing placed at the beginning of these two sections. We use a logo ( ), to indicate features on the overview drawings that are illustrated in greater detail in one of the "Steps" that follow the drawing.

Each "Step" starts off with our attempt to describe succinctly the generalized "Challenge" the builder faces at this stage in the process. This approach reflects our vision of this guide as a resource for the decision-making process you will step through on the way from your first fantasies to the first (of many) housewarming parties. At each major step in this process, the decision-making context will be unique. The "Challenge" facing you will have many possible solutions. Your shaping of the right solution for your unique situation will reflect many variables which only you can quantify or assess. Consider the following:

- your financial situation
- your timetable
- regional and micro-climatic factors and other physical characteristics of your building site
- your own availability as a worker and your skill level in various areas
- the availability of additional volunteer or paid labor at various skill levels
- the degree of your concerns about the sustainability, regional availability, healthiness and life-cycle costs of various materials
- the degree to which you use your time and labor to "buy" materials that have little or no monetary cost
- your personal comfort level for cost-cutting innovation/greater risk as opposed to typical overbuilding/greater security
- your aesthetic preferences and your willingness to pay for them.

The uniqueness of this combination of site, builder and building design suggests to us that a "cookbook" approach will not best serve you, our readers. There is no one "right" way to build a straw-bale structure or even to solve the problems to be faced at any given stage in the process. However, equipped with a modicum of common sense, a clear understanding of each "challenge", and of the unique properties of baled straw (or hay) as a building material and an array of options successfully used by earlier builders, we can all hope to shape a solution that is uniquely "right" for us.

Each "Challenge" is followed by an arguably chronological "Walk-Through" of the mini-steps that we envision taking you through this stage in the construction of the "model" building.

The drawings and text capsules that follow the "Walk-Through", provide coverage of options that have been used successfully by straw-bale builders. The logo ( ) is used to indicate features depicted in less detail in the overview drawing of our "model".

Throughout the guide we have tried to focus on those aspects of plan development and construction made unique by our decision to "build it with bales". Rather than repeating detailed information from other helpful sources, we provide literature citations (author, year of publication) for some of them. Complete citations are provided in the "Literature Cited" section near the end of the guide.

Following Part Two are three appendices that provide detailed information regarding sources and suppliers of materials and expertise. These listings are, of course, woefully incomplete. Let us know what/who you think should be added. Finally, mention or omission in these lists of any entity, person, product, or service does not imply endorsement of, or lack of confidence in, any of them. As always, buyer beware!

The final appendix provides excerpts from the Uniform Building Code and brief descriptions of the results of testing programs, to date, and how you can order detailed reports.

We end this opus with "Suggestions for Additional Reading and Viewing" and "Literature Cited". These provide practical resources in areas that we could not cover in detail while still keeping this guide brief and focused.

Roots

The saga of building human habitation with rectangular bales of hay or straw begins with the availability of mechanical devices to produce them. Hand-operated hay presses were patented in the United States before 1850, and by 1872 one could purchase a stationary, horse-powered baler. By about 1884, steam-powered balers were available, but the earlier horse-powered versions also continued to be used in the Great Plains at least through the 1920's.

We will probably never know any details of the first bale-walled building used to shelter human beings. It seems likely, however, that it's creator was a homesteader, recently arrived on the grasslands of the Great Plains and in desperate need of quick, cheap protection from a harsh climate. Although homesteading came to the Sandhills of Nebraska later than other parts of the Plains, it is here that we find the first, documented use of bales for building. The one-room schoolhouse, built in 1896 or 1897 near Scott's Bluff, survived only a few years before being eaten by cows.

The illustration, shown right (adapted from Welsch, 1973), provides a visual description of the technique apparently used in many of the early buildings. Of particular importance is the lack of any vertical posts to carry the weight of the light (generally hipped) roof, all of which was carried entirely by the bale walls. Use of the technique in Nebraska, most widespread from about 1915 to 1930, appears to have ended by 1940. Of the approximately seventy structures from this period documented by Welsch (1970), thirteen were known to still exist in 1993 and all but one of these (the oldest, from 1903) were still being lived in or used for storage.

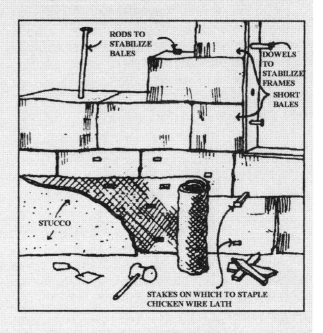

The Burke House, near Alliance, Nebraska, built in 1903.

Photo by Matts Myhrman

... and Revival

After its abandonment in Nebraska by 1940, the idea of bale construction wandered in search of folks motivated to build inexpensive, energy-efficient shelter. Rather than dying out, the method kept popping up in new locations as modern pioneers learned of it or re-invented it.

Welsch's 1973 article introduced the concept to a readership actively seeking alternatives, as did two articles that appeared in *Mother Earth News* (Doolittle, 1973 and McElderry, 1979). Another very important article, which appeared in the mainstream magazine *Fine Homebuilding* (Strang, 1983), described a small, post-and-beam studio designed and built by California architect Jon Hammond. In 1987, New Mexicans Steve and Nena MacDonald, two of the many individuals inspired by Strang's article, finally overcame their fears and built themselves a wonderful home that soon came to the attention of Matts Myhrman and Judy Knox. Inspired by Steve and Nena's home and philosophy, and building on the work of straw-bale pioneer David Bainbridge, Matts and Judy went on to set up Out On Bale, a straw-bale-construction education and resource center and begin publication of *The Last Straw*, the journal of straw-bale construction. Meanwhile, Steve continued to help/teach others to build with bales and, with his son, Orien, developed the straw-bale construction primer that this guide expands on.

Aided by extensive coverage in newspapers and magazines (*e.g.*, The New York Times, National Geographic) and on television, the revival of bale construction has generated pockets of almost religious fervor in locations around the world. As of April 1994, Out On Bale had documented more than 150 bale structures built since 1940, in Australia, Canada, Chile, Finland, France, Mexico and the United States. New buildings have been going up in a wide range of climates, faster than anyone is recording them. Most of these buildings are in rural areas, built without "benefit" of building codes or inspections. However, we know of with-permit, non-loadbearing, residential buildings in California, New Mexico and Arizona, and with-permit, loadbearing, non-residential structures in Arizona, Colorado, and Oregon.

Structural testing related to wind and seismic forces has been carried out in Arizona on walls built from 3-tie bales (Eisenberg, 1993). More is planned. Accurate measurements have also been made by Joe McCabe of the R-value of 3-tie baled straw (as summarized in *The Last Straw*, Summer 1993). Recent testing performed in New Mexico (SBCA, 1994) has established an R-value and a flame-spread rating for 2-tie baled straw (exposed and stuccoed). Additional testing included wall resistance to perpendicular and horizontal forces. More testing is being planned. We are optimistic that the results of such testing will lead to the routine acceptance of bales for loadbearing walls in residential structures.

Meanwhile, in April, 1994 in Tucson, Arizona, Jon Ruez and Hesh Fisk completed (for Matts and Judy) the first, with-permit, loadbearing, residential, bale structure and another such building is nearing completion in nearby Santa Cruz County. Negotiations have now been concluded that should soon lead to appending a "prescriptive standard" for loadbearing straw-bale construction to the building codes for Tucson and Pima County in Arizona, providing a model and precedent for other jurisdictions around the world.

The unique combination of environmental, socio-economic and natural resource issues facing our species as we approach the 21st century challenge us to expand the choices that will lead us toward sustainable systems. We see this legacy of bale construction, passed on to us by our homesteading ancestors, as one such choice, a beautiful baby that got thrown out with the bath water, but managed not to go down the drain.

Photo by Matts Myhrman

Virginia Carabelli's straw-bale home under construction near Santa Fe, New Mexico, spring 1991.

Questions and Answers

Q. What do North American builders mean by the word "bale"?

A. They usually mean a variously-sized, *rectangular* bundle of plant stems, held together by two or three ties of wire or baling twine and weighing from about 40 to 95 pounds (18 to 43 kg.). Such bales generally consist of "straw", the dry, dead stems that remain after the removal of seed from harvested cereal grains. This is an annually renewable, little-valued byproduct of cereal grain production and great quantities are available for baling in many parts of the world.

Q. What would constitute the ideal "construction grade" bale?

A. It would be very dry, free from seeds, well-compacted, consistent in size and shape, and have a length that is twice the width. More on this later.

Q. Are such bales locally produced and readily available anywhere?

A. Unfortunately not, but there is usually something grown within reasonable trucking distance that can be baled for building. However, in some such areas the only bales available are the big, round ones. We trust that as demand develops globally for "construction grade" bales, entreprenuering farmers will gladly meet the demand. Sources for bales can often be provided by agricultural extension agents, grain growers associations, tack and feed stores, zoos and race tracks. If you can work directly with a farmer, you have a better chance of getting bales that approach the ideal.

Q. What are loadbearing versus non-loadbearing straw-bale walls?

A. Loadbearing walls, such as one finds in the hip-roofed, historic Nebraska houses, carry a share of the roof weight. Non-loadbearing walls, either because of the roof shape or the presence of a complementary framework, carry none of the roof weight. More on this later.

Q. Is straw-bale construction, particularly when done with loadbearing walls (*i.e.*, Nebraska-style), inherently less costly?

A. A straw-bale house built in mainstream fashion, by a contractor using only paid labor, cannot cost significantly less than a frame or masonry house providing the same interior space. From the standpoint of a cold-blooded, profit-margin-driven cost estimator, this is just an exterior wall system. Since the cost (labor and materials) attributable to the exterior walls of modest homes generally accounts for only fifteen to twenty percent of the total project cost, use of straw bales to replace insulation and wood, metal or masonry can only affect this already small piece of the pie. The cost increases due to wider foundations and greater required roof area will offset some of these savings. Real savings begin when the eventual owner and friends provide the labor for the wall-raising, wall-surfacing and for interior finishing. Additional savings result from the use of recycled materials and those that cost little more than the owner's time (*e.g.*, salvaged lumber, locally available stone and earth). Further savings result wherever the owner-builder can substitute his/her own labor for paid labor or reduce costs by assisting a paid tradesperson. But even if a straw-bale house did end up costing as much as its counterpart, we believe it will still be a better house—quieter, more energy-efficient, more joyful to live in, and less costly to the planet's ecological systems.

Q. Okay, but what about termites?
A. A house built of baled straw is at far less risk than a wood-frame building, at least in North America and Canada, since virtually all the termites found there are specifically evolved to tunnel into and eat solid wood. Some builders do a standard chemical treatment, however, if only to protect wooden door and window frames and furniture. In areas where termites are a severe problem, a metal termite shield can also be included in the foundation design.

Q. Alright, but what about spontaneous combustion in a baled straw wall?
A. Spontaneous combustion can occur in large, tight stacks of **hay**, baled while still too green and wet. However, we have been able to document no case of this occuring with straw bales stacked in a wall.

Q. Yeah, but what about fire?
A. As long as the bales are covered with stucco, mud plaster, gypsum plaster, aluminum siding or sheetrock, a bale building will be extremely fire-resistant. Even an unprotected straw-bale wall performed satisfactorily in tests performed by a certified testing lab in New Mexico (SBCA, 1994).

Q. Then what about vermin (*i.e.*, rodents and insects)? Do the bales need any special chemical treatment to protect against them?
A. As in a frame structure, the secret lies in denying unwanted critters a way to get in and out of your walls. Build so as to isolate the bales (including the tops of the walls) and then regularly check and repair the exterior and interior wall surfacing. A few modern builders have used bales having lime incorporated into them or have dipped or sprayed the bales using a lime slurry or borate solution. Such measures may provide an extra level of insurance if maintenance of the wall surfacings is poor.

Q. Is straw-bale construction suitable for all climates?
A. The only serious enemy of straw is prolonged exposure to water in liquid form, since with sufficient moisture present, fungi can break down the woody stems. High humidity, by itself, does not appear to be a problem, but few historic examples exist from areas characterized by consistently high relative humidity. Walls continuously exposed to high humidity from within or without could experience condensation within the walls during periods of extremely cold temperatures. In such situations, moisture barriers (in reality, air movement barriers) are sometimes used on the inner surface of the exterior walls, often in conjunction with an air exchanger to prevent interior air quality from becoming unhealthy. High rainfall can be dealt with by proper design and detailing (*e.g.*, adequate roof overhangs) and regular maintenance of the roof and wall surfaces. Since thick bale walls are highly insulative, the ideal climate for straw-bale construction may be arid or semi-arid, with hot summers and cold winters, but successful examples exist in a wide range of climates.

Q. What about durability/longevity?
A. The evidence provided by existing hay

and straw-bale structures built by Great Plains homesteaders as early as 1903 is irrefutable - bale houses, if properly built and maintained, can have a useful lifespan of at least 90 years, even in areas where high winds are common. Specialists in earthquake-resistant design have predicted that structures with properly pinned bale walls will be unusually resistant to collapse during earthquake-generated motion.

Q. Can the technique be used in places where a building code is enforced?

A. In the rural counties of many mid-western and western states, only people living in or near the largest communities are required to obtain building permits and undergo inspections that confirm compliance with the Uniform Building Code (or one of the other codes), as adapted and amended by that specific jurisdiction. In these areas, homebuilders are often still required to use a loadbearing framework to support the roof, and can use the bales only as insulative wrap or infill. Some structural testing has already been completed in Arizona and New Mexico to provide quantitative evidence that bales can be safely used in loadbearing walls (Eisenberg, 1993 and SBCA, 1994). On this basis, permits have been obtained in Tucson and in Santa Cruz County (all in southern Arizona) to build small, residential structures with loadbearing walls.

Q. Does the use of bales impose limitations on the building design?

A. If a loadbearing framework is used to carry the roof weight, the limitations are very few. One could conceivably build a multi-storied building with straw-bale infill or wrap a huge single-story building with non-loadbearing bale walls.

However, if one wishes to use the walls to carry the roof weight, the unique properties or idiosyncracies of bales as loadbearing units must be given serious consideration. The compressible nature of bales will suggest reasonable limits on the following: 1) the height of walls (generally, a maximum of seven courses for either 2- or 3-tie bales laid flat); 2) the number, width and location of openings; 3) the length of trusses or rafters; and, 4) the weight of roofing systems. Where more space is required than can be comfortably provided by a single-story square or rectangle (of acceptable length), builders have turned to "bent rectangles" (e.g., L-shapes, U-shapes or designs with fully surrounded courtyards). Another strategy is to create additional living space under a "sheltering roof" (e.g., gable, gambrel or hip).

A wide variety of roof styles can be used, the least desirable perhaps being a "flat" roof surrounded on all four sides by vertical parapet walls. Creating this kind of "bathtub" above straw-bale walls is asking for trouble.

Whether or not they are loadbearing, bale walls are invariably thicker than those resulting from standard frame or masonry construction. In feeling, they more resemble double adobe or rammed earth walls. Unlike earth walls, they cannot practically be left permanently exposed, but a wide choice of coverings can be used (e.g., cement- or lime-based stuccos, gunite, adobe mud plasters, metal or vinyl siding, wood paneling or sheathing, gypsum-based panels [e.g., sheetrock, drywall] and gypsum-based plasters). Many bale buildings have exterior walls covered differently on the inside and outside.

Q. Thick bale walls may look and feel nice and provide great insulation, but don't they require roofing too much unusable interior space?

A. One of the unavoidable trade-offs of thick walls, be they super-insulated double frame, rock, rammed earth, adobe or straw bale, is that to provide the same usable interior space as with thin walls, you must widen the foundation/slab and create more roof surface area. As an example, imagine a 1500 square-foot house (based on exterior dimensions) with seven-inch thick walls (2"X 6") frame with fiberglass insulation plus one inch of extruded polystyrene foam board) providing a total R-value of about 28. To maintain the same interior space with 24-inch thick plastered straw-bale walls (about R-50), we would need to increase the square footage of the foundation by 15% (see graph, next page). Is this too much for what we'd get in return, including some fine interior widow-sill spaces? You're the judge.

Q. What about obtaining construction insurance and a building loan convertible to a mortgage? Do such houses have normal resale value and can potential buyers get financing?

A. The early straw-bale houses were uninsured, pay-as-you-build structures. However, some recent builders have obtained construction-phase insurance convertible to homeowner's insurance upon completion of the project. Farmer's Insurance Company presently provides insurance for two non-loadbearing homes under the category "frame-stucco". Companies will undoubtably be more reluctant to insure houses that lack a loadbearing framework, but a with-permit guesthouse with loadbearing walls has been approved for coverage under an existing homeowner's policy by the Continental Insurance Co. of New York.

The resale value of straw-bale homes is difficult to determine since, at least to our knowledge, no modern straw-bale has ever been put on the market. The growing popularity of the technique in the Santa Fe, New Mexico, area will guarantee public exposure, enhance its desirability and encourage greater acceptability with the insurers and lenders.

Q. Will a straw-bale house cost less to heat and cool than a typical frame or masonry house, assuming comparable interior size, shape, ceiling insulation and solar orientation?

A. Since typical construction seldom provides wall-system R-values greater than 20, a well-built, straw-bale house with walls providing R-values of from R-40 to R-50 (depending on surface coverings, density of bales, thickness of walls, etc.) will obviously cost less to heat and cool than a typically built home. These energy savings, which will be proportionally greater for smaller designs than for larger ones, will accrue to the owner month after month for the lifespan of the building.

Q. Since straw-bales are a relatively low-mass material, will they work well in a passive solar design?

A. The major physical components of an ideal passive solar design would include adequate thermal mass (to store and release heat on a 24-hour cycle) and an insulating exterior wrap to reduce heat loss to the outside. In straw-bale construction, proper placement of high mass materials like stucco, mud plaster, brick, concrete, tile, adobe or rammed earth in the interior of the structure provide the needed thermal mass, while the thick, highly insulative, straw-bale walls greatly reduce heat loss by conduction. Straw bales on the outside, earth on the inside—we win, the planet wins.

Q. Strictly from the standpoint of maximizing the advantages and minimizing the disadvantages of straw-bale exterior walls, is there an ideal size for a simple, rectangular building?

A. This is obviously a very narrow way of looking at how big or small a building should be, but several factors point to dimensions that will provide an interior useable space of about 1200 square feet. This size building has a ceiling area approximately equal to the internal surface area of the outside walls, so the impressive R-value you get by stacking the bales is not overshadowed by ceiling area that one needs to create (with lumber) and insulate. At the same time, by having an interior as large as 1200 square feet, we have reached the point where the square footage of the "pad" needed is only 16.5% greater than that needed for a house using an R-28, 2 X 6 frame system. Such a straw-bale house would need a pad providing just under 1500 square feet (see below).

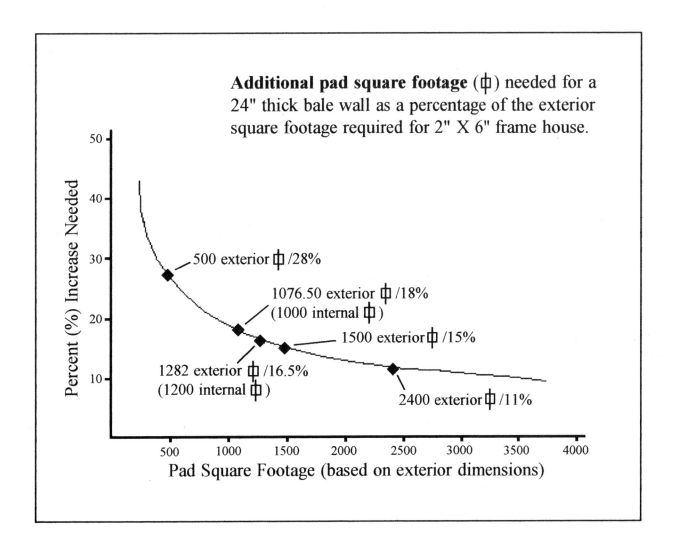

Additional pad square footage (⊡) needed for a 24" thick bale wall as a percentage of the exterior square footage required for 2" X 6" frame house.

500 exterior ⊡ /28%

1076.50 exterior ⊡ /18% (1000 internal ⊡)

1500 exterior ⊡ /15%

1282 exterior ⊡ /16.5% (1200 internal ⊡)

2400 exterior ⊡ /11%

Percent (%) Increase Needed

Pad Square Footage (based on exterior dimensions)

Part
One
Before You Build

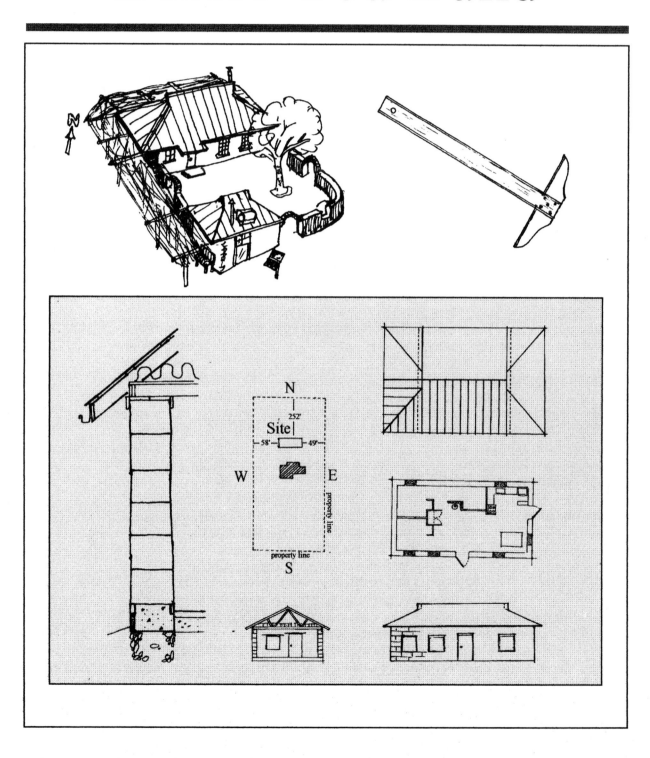

A Bale is a Bale is a ...

Hay versus Straw

The term "hay" is used to describe the material which results from cutting certain plants while still green and allowing them to partially dry before removing them from the field. Stored in stacks or bales until needed, this nutritive product is fed to animals. Contrast this with "straw", the dry, dead stems of plants, generally cereal grains, that is sometimes removed from the field after the seed heads have been harvested. The majority of this low-value, nutrition-poor byproduct is burned or tilled into the soil -- only a small percentage of that which is available is baled. Although baled meadow hay has been used by both historic and a few modern builders, straw is the preferred (and generally cheaper) material.

Bale Options

Bales come in a variety of sizes and shapes, but those most commonly used for building are the small rectangular bales. These come with either two or three ties, and the ties may be wire, polypropylene twine or natural fiber

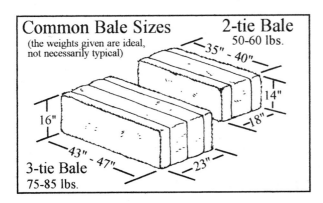

Common Bale Sizes
(the weights given are ideal, not necessarily typical)

2-tie Bale
50-60 lbs.
35" - 40"
14"
18"

16"
43" - 47"
3-tie Bale
75-85 lbs.
23"

twine. The vital statistics are given below. Builders have generally favored polypropylene twine, because it cannot rust, but wire runs a close second. Natural fiber twine is considered a final resort due to its low tensile strength and susceptability to rot. Builders have commonly used both two- and three-tie bales in "post-and-beam" applications, and have used them laid both "flat" and "on edge". For designs with loadbearing walls, most builders favor the more-compact, wider, three-tie bales laid "flat". Successful examples do exist, however, of structures with loadbearing, two-tie bales laid "flat", and loadbearing, three-string bales laid "on edge".

The Ideal, Building-Grade Bale

This hypothetical super-bale would be:

• dry – the drier the better. At a moisture content below about 20% (calculated as a percentage of the total weight of the bale), virtually none of the species of fungal spores commonly present in straw can reproduce and cause the straw to break down.

• free from seed heads that would encourage rodents to inhabit the walls should the wall surface "skins" not be properly maintained.

• about twice as long as it is wide. Such bales, when laid "flat", will lay up with a true running bond, where each vertical joint between two adjacent bales in a course will fall at the mid-point of the bales above and below the joint.

• consistent in size, shape and degree of compaction with its neighbors. Such bales would make it easier to build straight, relatively smooth-surfaced walls of uniform height. This, in turn, minimizes the amount of bale-tweaking needed to remove excessive irregularity. It also decreases the amount of stucco or mud plaster, if these are being used, that will be needed to achieve the desired amount of wall smoothness.

• sufficiently compact for its intended use. This proves to be easier to suggest than to provide standards for. Until some inexpensive, easy-to-construct, standard device has been adopted to physically measure the degree of compaction of baled plant stems, we're stuck with using density (loosely defined as weight per unit volume) as an easily measured substitute for degree of compaction.

For non-loadbearing use, the degree of compaction is much less critical, since the bales are braced against forces perpendicular to the wall surfaces by the roof-bearing framework. The proposed code for the State of New Mexico for non-loadbearing construction requires only that the bale can be picked up by one string without deforming. If the bale walls will be carrying the roof load, the degree of compaction will affect the stiffness of the pinned walls and their resistance to wind and seismic loads. It will also influence the total amount of wall compaction resulting from a given load-per-lineal foot on the top of the wall and the time required for this compaction to be completed.

Which is all fine and good, but still leaves us needing a way to easily determine the "calculated average density" of our bales to see if it exceeds some accepted, minimum value. The standard procedure has been to weigh a given bale and to then estimate, using a measuring tape, the dimensions of an "envelope rectangle" that would snugly enclose that bale. Each dimension in inches and eighths (e.g., 46-3/8"), needs then to be converted to inches and decimal inches. To do this, divide the numerator of the fractional inches by the denominator (e.g., 3 divided by 8 equals 0.375") and add this to the whole inches (e.g., 46.375"). Now multiply the converted length, width and height and divide the result by 1728 (the number of cubic inches in a cubic foot). Divide this result into the weight of the bale (in pounds) to obtain the "calculated density" in pounds per cubic foot.

The accuracy of this result depends, unfortunately, on the assumption that the bale is bone dry. The presence of moisture (*i.e.*, liquid water) in the bale will give a falsely high result. We have few choices in how to deal with this wild card.

Moisture content is normally expressed as a percentage of the total weight of the "damp" bale. An 80 pound bale in which 16 pounds was due to water would have a moisture content of 20% (16 divided by 80 = .20). Digital moisture meters, with a probe that is stuck into the bale, are available (although expensive). They are calibrated for alfalfa hay, not straw, and cannot read accurately below about 13%. However, this is below the generally accepted upper limit for safe use in building (*i.e.*, 20%) . At moisture contents above 20%, many of the fungal spores naturally present in straw can reproduce and break down the cellulose that constitutes straw. Simple lab procedures performed on samples taken from bales can determine the moisture content very accurately, but require destroying the bales and are time-consuming and expensive.

What is needed, but not yet available, is a method for determining the degree of compaction that is independent of the moisture content and that can be performed quickly, on the building site, with an inexpensive device. Send us your ideas!

Sources of Bales

In areas where cereal grains (*e.g.*, wheat, oats, barley, rye and rice) are produced, it is often possible to buy bales cheaply in the fields or from stacks located beside the fields. Farmers will sometimes load and deliver, but transportation is generally provided by independent truckers. The cost per bale, bought retail in small quantities at a "feed store", is often significantly higher than the price that can be negotiated, through the feed store or the producer, for a larger quantity. Other potential sources of information on bale suppliers include state agricultural agencies, county agricultural agents, race tracks and zoos. See *Appendix One* for access information on some major suppliers.

Uses of Straw

A non-inclusive list would include the following:
- agriculture and gardening (disked in as a soil loosener; on the surface as mulch)
- crafts (especially in England and Scandinavia)
- livestock bedding

- roughage in cattle feedlot mixtures
- erosion control (for small check-dams and in hydro-seeding)
- manufacture of paper and pressed-straw board
- construction (insulation of concrete slabs during curing; in straw-bale walls and as ceiling insulation)

Bale Composition

The straws of the common cereal grains are very similar in chemical composition to each other and to the common soft woods. They all consist mainly of cellulose, hemi-cellulose and lignin. It is far more important that the bales be dry and compact than that they be wheat rather than oats. Even sudan grass, bean stalks, and green tumbleweeds have been baled and used successfully for building.

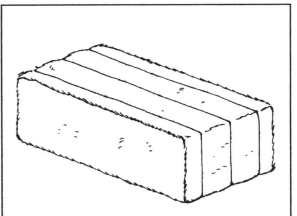

For designs with *loadbearing walls*, most builders favor the wide and compact three-tie straw bales, laid flat.

How Many Bales?

If you have completed your plans, including detailed wall "elevations" (i.e., vertical wall maps), you can determine very exactly how many bales you will need. The determination is often done as if there are no doors and windows, to insure that there will be some extra bales for temporary seating and for building ramps or to support scaffolding. For an initial estimate, however, use the graph below.

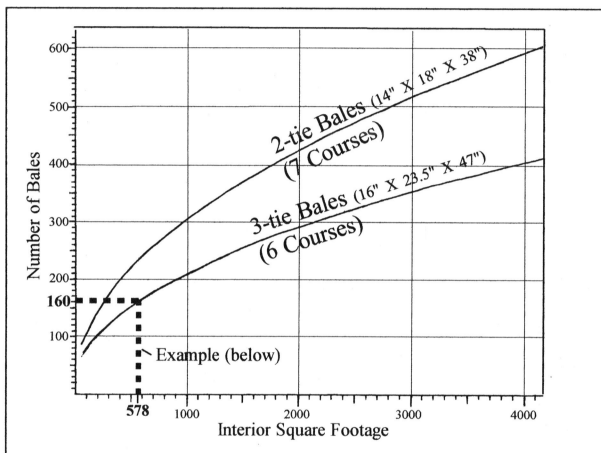

Estimating the number of bales (2-tie and 3-tie) needed to build your home. The horizontal axis is the square footage of interior space; the vertical axis the number of bales needed. We assume no openings in the walls. In the **example** shown, the interior square footage of our planned building is 578 square feet if 3-tie bales are used. Coming straight up from 578 on the horizontal axis until we intersect with the 3-tie curve, we can then move horizontally to the left until we intersect the vertical axis at 160, the number of bales needed.

Three Basic Approaches

Loadbearing

In nearly all cases, the roof weight of the historic Nebraska structures was carried entirely by the bale walls. Many of these buildings were square or modestly rectangular, with lightly-framed hip roofs that distributed the weight evenly, or nearly so, onto all four walls. The advantages of this approach include:

- the greater ease of design and construction.
- savings of time, money, labor and materials, since no roof-bearing framework is needed.
- distribution of the roof and wall weight evenly along the foundation under the loadbearing walls.

The disadvantages include:

- certain design constraints, including the need to avoid very heavy roof systems.
- the need for dense, uniform bales, laid "flat".
- the need to wait an indeterminate period of time (generally, 3 to 10 weeks) for the bales to compress in response to the weight of the roof/insulation/ceiling system, unless your tie-down system can pull the roof plate / bond beam down mechanically.
- the possibility that very heavy live loads (wet snow, herds of elephants, etc.) could cause the wall surfacing materials to buckle outward.

Non-Loadbearing

Many of the modern bale-walled structures have been built using an arrangement of vertical elements (generically called posts) and horizontal elements (beams) to carry the entire weight of the roof/insulation/ceiling system. The bale walls carry only their own weight, providing insulation and the matrix onto which surfacing materials (e.g., stucco, mud plaster, siding) are attached. Typical frameworks have consisted of various combinations of 2"x4" and other dimension lumber, glue-lam beams, rough cut timber posts and beams, peeled logs, metal elements, and concrete columns and bond beams.

The advantages include:

- greater familiarity, and therefore acceptability, to building officials, lenders and insurers.
- provision, by the roofed framework, of a dry place for storing materials, including bales, which enables flexible scheduling and working even when it's raining.
- the possibility, since the framework is non-compressive, of surfacing the walls as soon as they are up.
- the possibility of using the typically less-dense, two-tie bales or less-dense-than-normal three-tie bales, and laying the bales "on edge".
- A reduction in constraints on the size, number and placement of openings.
- freedom from certain other design constraints (e.g., length of unbuttressed walls, roof weight).

The disadvantages include:

- the expenditure of extra time, money, labor and materials to create a second, loadbearing system if the bales themselves could have handled the load.
- the need to create a more complex

foundation system that can carry both the bale walls and the concentrated loads transferred to it by the vertical "posts".

Hybrid Combinations

Some modern structures have been designed to share the roof load between loadbearing straw-bale walls and other elements (e.g., frameworks or central columns). Since the former compress, while the latter do not, particular care must be taken in such designs. The hybrid approach allows for larger or two-story structures that still have exterior, loadbearing walls and for south-facing (in the northern hemisphere) walls, consisting largely of glass.

Native American grass house (drawn from photograph in *Nabokov and Easton 1989:144*)

Developing a Plan

A Different Way Of Building

Straw-bale construction encourages us to explore a different philosophy of building, one which includes concepts like those listed below:

• use passive heating and cooling systems to the extent possible.

• design to enhance eventual expansion, but building now only what is enough for now.

• build accretionally, with final inspections as each major stage is completed. For example, complete the basic core, with kitchen, bathroom and a multi-purpose living space. Move legally into this and then add a bedroom. Get this inspected and start using it. Future additions might add a master bedroom and bath or an office space.

• transition gradually from a small, actually mobile, trailer, to the trailer nestled up against a straw-bale "great room", to the trailer totally within (except for an exit hole) an enlarged straw-bale building, to the trailer gone on to spawn another transition (see drawing on page 63).

• keep the design simple, the size small (keep asking yourself if you've drawn more than enough), the spaces multi-functional, the partitions easily movable, the storage inventive (utilizing otherwise dead spaces). In temperate climates, consider a fully-climatized core, with zones around it that become increasingly less enclosed as one moves out.

• use "green" and healthy materials where possible (see Bower 1993; Breecher 1992).

• do all or part of the building oneself (a family effort).

• pay for the building as you go, as you can. Imagine no mortgage payments and, particularly, no interest payments. That *would* be another galaxy, and a kinder, gentler one at that.

You will find such concepts reinforced and elaborated on by Anderson and Wells (1993), Alexander (1979), Kern (1975), Rybezynski (1989), and Thoreau (1950).

Preliminary Conceptual Design

After coming up with some loose, informal, preliminary sketches, but before spending time developing detailed plans, you may want to initially consider some relevant, generic issues.

Very early in the process, while you're checking on the availability of good bales, explore the code situation for your state, county or municipality. Should a building code apply to the type of building you propose, you must then decide whether to do the building legally, or "bootleg" it. Before choosing the latter, consider carefully the consequences of being discovered.

In working with your building department, it will help to provide them with information about this technique as early as possible. Bend over backwards not to develop an adversarial relationship with them. This doesn't mean being a wimp and letting them bluff you into thinking you can't do something that you believe is legal. It does mean doing your homework (get and study the appropriate zoning descriptions and code sections, see *Appendix Three*) and making them aware, in a non-threatening way, that you have done so. Assure them that you want to work with them to do something that is safe and legal, and provide testing data as backup. Put them in touch with building officials in jurisdictions that have already issued permits. If necessary, utilize the existing appeals process. Good-natured, informed perseverance is a formidable tool.

Good resources for working with building codes and officials are Eisenberg (1995) and Hageman (1991). Your local building department will have a copy of the code they use, for you to read on the premises.

Structural Implications of Openings in Bale Walls

When we stack and pin bale walls, we create a sort of "fabric", whose strength and stiffness is strongest when no openings are left in it. Skylights don't affect this "fabric", but doors and windows will. A general rule followed by most modern bale builders is to place no opening closer than one and a half bale lengths to any corner or to another opening. Wide openings require a stronger, heavier lintel to bridge the opening or a beefier, loadbearing frame to carry the roof and/or wall weight sitting above it, so openings are often kept narrow (see *The Loadbearing Option, Step 2*, for details). Windows placed higher in a wall are shaded more of the year by a modest roof overhang. If the climate

dictates a large amount of south-facing glass, it may be wise to consider using frame or post-and-beam in the south wall (see *Part 2, The Hybrid Option*).

Idiosyncracies of Bales as a Construction Material

Structural engineers involved with the revival tell us that bales are unlike any building material that they normally encounter. The basic technique resembles masonry, but masonry units (adobe blocks, fired clay bricks, concrete blocks, etc.) are brittle, non-compressive and fail catastrophically when loaded past their limit. Wood frame construction has some inherent flexibility, but is essentially non-compressive under vertical loading until failure occurs. Bale walls are flexible, compressive and relatively elastic, responding to loading by gradual deformation rather than sudden, brittle failure.

The structural idiosyncracies of bales are only the first in a long list of things about them that you might consider as you begin the design/build process:

• In a given area, bales are usually produced during a short period and the supply for the next year is fixed at this point. An additional concern is the regional availability of compact, three-tie bales versus the typically looser, less compact, two- tie bales. Rather than have three-tie bales brought in from great distance, you may decide to have customized, denser-than-normal, two-tie bales produced locally to meet your needs.

• A load of bales, even if all have come from the same field and from the same piece of baling equipment (baler) with the same operator, may show considerable variability in dimensions (primarily length), degree of compaction and moisture content.

• All wall systems have combined dead (*e.g*, roof/ceiling) and live (*e.g.*, snow) loads that cannot safely be exceeded. Calculations can be made, using data from engineering testing on bale walls, to determine whether the maximum predicted roof load exceeds acceptable limits. Even within these limits, a heavier roof load will result in more compression (and perhaps a longer settling period). Get help from an architect or engineer if you have any doubt about your ability to do these calculations or interpret the results. Or, in areas not experiencing snow loads, trust the historical record, keep the building small and the roof light, and you'll probably do just fine.

• A variety of materials are used to tie bales, including rustable, non-galvanized wire, polypropylene twine and a variety of natural fiber twines. Our general concerns must be that the tie material be strong, resistant to rust or rot where exposed repeatedly to damp stucco, and not attractive to rodents. Even if the bales are being laid flat (so that the ties are within the wall), the ties will be exposed where bale ends form a corner and, to a lesser extent, where bales butt against a door or window frame. If wire-tied bales are being used, backup ties of galvanized wire or polypropylene twine can be added before bales are laid at such locations. In general, polypropylene twine is favored over regular baling wire by many builders and fiber-tied bales are avoided.

• If simplicity and speed of stacking is a major goal, the design should involve only full and half bales. This is possible only if the bales are about twice as long as they are wide and if all openings are some whole number multiple of half of the effective bale length (more on this later).

• Under normal circumstances, the only enemies that the bales have are the ever-present fungi. Even these are harmless if the bales remain dry, but in the presence of sufficient, liquid moisture they can gradually break down the straw. The irrevocable contract that you make with your bale walls is that you will protect them from water (*i.e.*, liquid water).

• To our knowledge, all loadbearing, bale-wall designs to date have been limited to a single story, and the bales have, almost without exception, been laid flat. Some single-story designs have used a sheltering gable, gambrel or hipped roof to provide additional living space. Based on the engineering testing now available, however, some intrepid designers are developing designs for round, two-story structures incorporating a rigid, central post or column that carries part of the roof and floor loads and provides additional resistance to the wind loads resulting from the higher walls (see pg. 63.).

• Finally, bales are carve-able (for wall niches), bendable (for curved walls), and dividable (for custom-sized and shaped bales). You don't *need* power tools to work with them, so the building site can be wonderfully quiet.

Moisture Protection Strategies

The matter of how to protect the straw from liquid moisture that may reach it in a variety of ways should not be treated lightly. Even formulating the right questions is difficult, and the answers are very specific to the local climate and even micro-climate. In any case, it makes sense to prevent water from reaching the bottom course of bales from below or from the exterior and to provide a waterproof drape at window sills and at the top of all walls. A review of the "building science" literature on moisture protection reveals significant disagreement among the "experts", and not just on the picky details. Wood, which is chemically similar to straw, is also subject to water damage. This means that long-time, conventional builders in your area can provide relevant advice. The written resources range from popular magazine articles (with over-simplified, cookbook answers) to scholarly texts that will panic and confuse all but the engineers among us (and sometimes them, too). We can recommend Gibson (1994) for a good overview and Lstiburek and Carmody (1993) for detailed coverage.

Mechanical and Electrical Systems, Etc.

Before starting to develop a floor plan, you would be well-advised to decide how you will deal with your needs for the following:

- water (water harvesting [Pacey and Cullis 1986]? well? etc.)
- transportation (interior space for vehicles?)
- electricity (grid? stand-alone PV? wind? water? (see Strong 1994, Potts 1993)
- gas or LPG(butane or propane)
- disposal/usage of "wastewater" (composting toilets? artificial wetlands?
- space heating (passive/active mix? radiant floor [Adelman 1984, Luttrell 1985]? wood stove? etc.)

- food production (*e.g.*, attached sunspace / growspace)
- cooling (venting? evaporative cooler? cooling tower?) (see Lecher 1991, and Cook 1989).
- clothes washing and drying
- cooking

Such decisions may heavily influence certain aspects of the floor plan(s) and roof shape. A decision to rely heavily on roof-top water harvesting could, for example, lead to incorporating the storage container(s) into the structure as a loadbearing element (or elements) and/or as thermal mass. It might also suggest including generous roof overhangs or porches to increase the harvesting surface and a simple roof design that can be easily equipped with gutters.

Developing a Building Plan

Here's where you finally get down to the nitty-gritty of developing the building plan(s). You aren't likely to forget to provide spaces for cooking, eating, excreting, bathing, sleeping, lovemaking, socializing and relaxing. However, don't forget to design in space for traffic and air flow, various kinds of storage, a home office, your mechanical systems (space and water heating and cooling/ventilation, appliances [esp. washer/dryer], etc.). As mentioned earlier, consider building additional small buildings later as needs change or letting a single structure grow over time by pre-planned additions.

Even if you eventually plan to sell your house, don't let "resale" considerations bludgeon you into creating one-plan-fits-all, generic blah. Let your instincts and creativity be reflected in a design that delights you and yours, while not making it so personal a "glove" that it can comfortably serve none but you.

Sources that we have found helpful for the design process are Alexander (1977), Day (1990), Jacobson (1990), Taylor (1983), Ballard (1987), Talcott (1986), Jones (1987), Connell (1993), and Cecchettini *et al.* (1989).

Site Selection Considerations

On a small lot, you may have little latitude in positioning your building. Given a larger piece of land, careful study of the whole piece will probably reveal several sites better used for building than for something else. One can then try measuring each site against this list of some of the characteristics of the "ideal" building site for a floor-on-earth, straw-bale building:

- the views are attractive to you (could mean vast or restricted)
- access to the site is reasonable
- there is a reasonably flat area big enough for the building
- if the site has a slope, it is generally toward the south
- the drainage pattern will not present any unplanned-for difficulties
- the position of the site in the general landscape will ameliorate the least attractive aspects of the climate rather than accentuate them
- winter sun, for passive solar gain, reaches the site
- the geology of site is such as to minimize problems and expenses related to site preparation and foundation design and construction

Although none of your sites may exhibit all the desired features, this exercise will enable you to compare them and develop a ranking that reflects how strongly you feel about the various "ideal" characteristics. Valuable aids for this process include McHarg (1969), Mollison and Slay (1988), Lynch and Hack (1984), and Erley and Jaffe (1979).

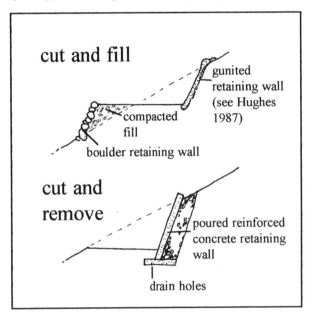

Site Preparation

This step involves whatever modification of the site is necessary in order to be able to lay out the building and create the foundations. For a flat site this may be as simple as scraping the surface to remove vegetation, loose soil and roots. If the site slopes, varying amounts of cutting and filling can be done by hand, or with machinery, to create a level pad large enough to put the structure on.

Any fill soil must be adequately compacted as it is put in place to insure that the material will not later settle under the weight of the building (see Monahan 1986). Steep cut and fill slopes will need retention and/or stabilization (see Erickson 1989). If the bale structure is to sit on a wood deck supported by posts or piers (a common response to steeply sloped sites), one need only completely clear the area where excavation for the posts will be done.

It makes sense to try to finish the site preparation before finalizing the design, because problems encountered during the site preparation may suggest major modifications in shape of the building or in the foundation system initially chosen.

Finalizing The Design

If the approximate upper limit on bale length is known, for the specific bales to be used, one can finalize the foundation dimensions using the chosen bale layout and this length. Since few bales will approach this upper length limit, occasional flakes of loose straw will have to be used to fill small gaps as the bales are laid up. This is quickly and easily done and does not significantly weaken the walls.

Another common approach, if one is lucky enough to already have one's bales, is to determine the "effective bale length". This is done by arranging ten, randomly selected bales butted snugly end to end in a straight line. With short boards held vertically against the ends of the arrangement, the distance between the inside surfaces of the two boards is measured and divided by ten. The resulting number, in decimal inches or meters, is the effective bale length. Seasoned builders often add a quarter of an inch to provide a little cushion. The half "effective bale length" can

also be used to finalize the width of openings. Bales can also be stacked in wall-high vertical columns to determine the "effective bale height". This figure is useful in finalizing door and window frame heights.

At this point one should prepare scaled drawings (where a certain distance on the drawing equals a fixed distance on the ground) of the bale layout for courses one and two. Except for the presence of window and doorway openings, all odd-numbered courses will be repeats of course one and all even-numbered courses will mirror course two. A scale of one-quarter inch equals one foot is commonly used in countries not using the metric system.

Architects and building professionals are trained to effectively use two-dimensional drawings to represent three-dimensional buildings. For the rest of us ordinary mortals, models can reveal a world of problems and solutions. Nearly true-scale micro-bales can be purchased from craft supply stores. Exactly true-scale bales can be handmade from 1/2" or 1" styrofoam insulation board (the high-density variety cuts cleaner) or wood. These enable one to build a scale model of one's building on the kitchen table and get most of the glitches out of the design

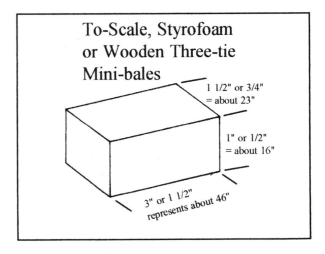

To-Scale, Styrofoam or Wooden Three-tie Mini-bales

1 1/2" or 3/4" = about 23"

1" or 1/2" = about 16"

3" or 1 1/2" represents about 46"

before things get too real (see Feirer 1986).

Even without a model, we can now proceed to prepare a vertical wall-map (*i.e.*, an "elevation") of each wall, showing the placement of each bale and half bale, and of all door and window frames and lintels (if any), as viewed from the outside (don't ever stand inside the building when using one of these maps to position a niche or opening). These maps are invaluable during the wallraising and should be posted in front of each wall for frequent, convenient reference.

Before moving on to creating your final working drawings, review all decisions about plumbing, mechanical, and electrical systems to ensure that any changes made along the way are still accommodated by the floor plan.

Now prepare, or have prepared for you (by an architect or construction draftsperson), detailed, scaled working drawings that will help you keep track of your decisions. We recommend that you have, as a minimum, the following:

- **a site plan**, showing how the building fits on the site, along with any easements, power sources, underground pipes, etc.

- **a floor plan**, showing interior partitions, window and door openings, stairways, porch extensions, etc.

- **a foundation plan**, showing locations of foundation bolts, rebar pins, eye-bolts, etc.

- **a cross-section of the foundation system**, showing reinforcement bars ("rebar"), perimeter insulation, floor design, etc.

- **a roof-framing plan**

- **cross-sections of the wall system** itself, and at typical doors and windows

- **"elevations" of each wall** with detailed sketches of each door and window frame

- **a plumbing and mechanical plan**

- **an electrical plan**

These same drawings will constitute much of the package that you will have to provide if you are applying for a building permit. Consult with your local building officials regarding their specific requirements for a permit application. Useful resources for this process include Syvanen (1982), Curran (1979), Weidhaas (1989), and Spence (1993).

Preparing a Materials List

At this point, many builders do a "takeoff", *i.e.*, prepare a comprehensive list of needed materials, doors and windows, hardware, fasteners, etc. (see Alfano 1985). To prepare such a list, sit down with a detail-compulsive friend and start with the very first step (usually, building layout). Step through the whole project, one task at a time, and figure out everything you need to buy, borrow or rent to support each task.

Now, review your shopping list. Well in advance of when you will need them on the building site, order any materials not locally available, right off the shelf. Prepare, as necessary, to store these and other materials on site with whatever protection they require. Keep your bales off the ground (shipping pallets work fine) and under roof or well-covered with plastic and/or a tarp. Crown the top of the stack to encourage rain to run off rather than puddle and leak.

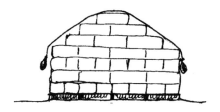

At long last, you should be able to send out invitations for your wall-raising and to tentatively schedule the other activities in the sequence that will lead to a finished building. Veteran builders factor in Murphy's Law, delays in the arrival of materials, bad weather, that unannounced three-week visit by your in-laws, the flu, etc. For each major work effort, line out the person-power and equipment that you'll need, and figure out how you're going to get it. Don't schedule move-in or the house-warming party quite yet. You'll have plenty of time to do this later as light begins to appear at the end of the tunnel.

And now, into the breach we go!

Photo by David Noble, Photographer, Santa Fe, NM

Part
Two
Building At Last!

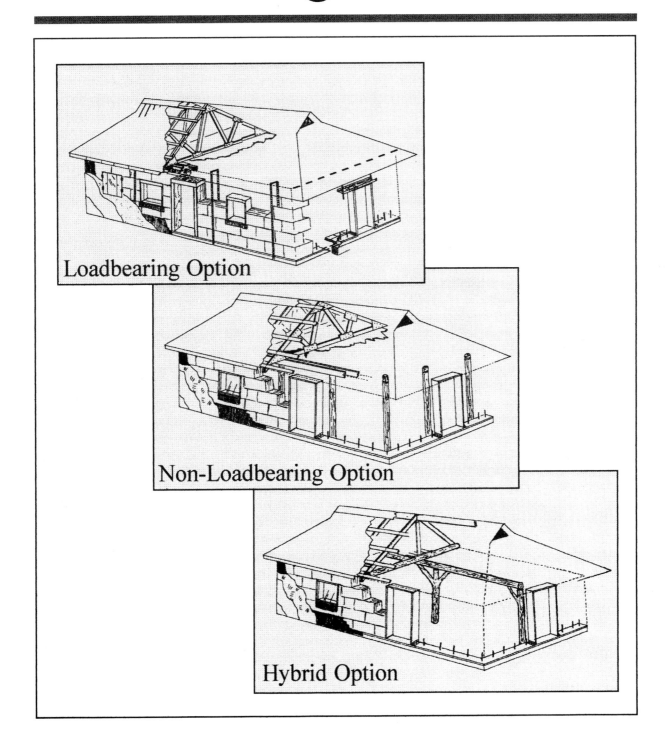

Loadbearing Option

Non-Loadbearing Option

Hybrid Option

The Loadbearing Option

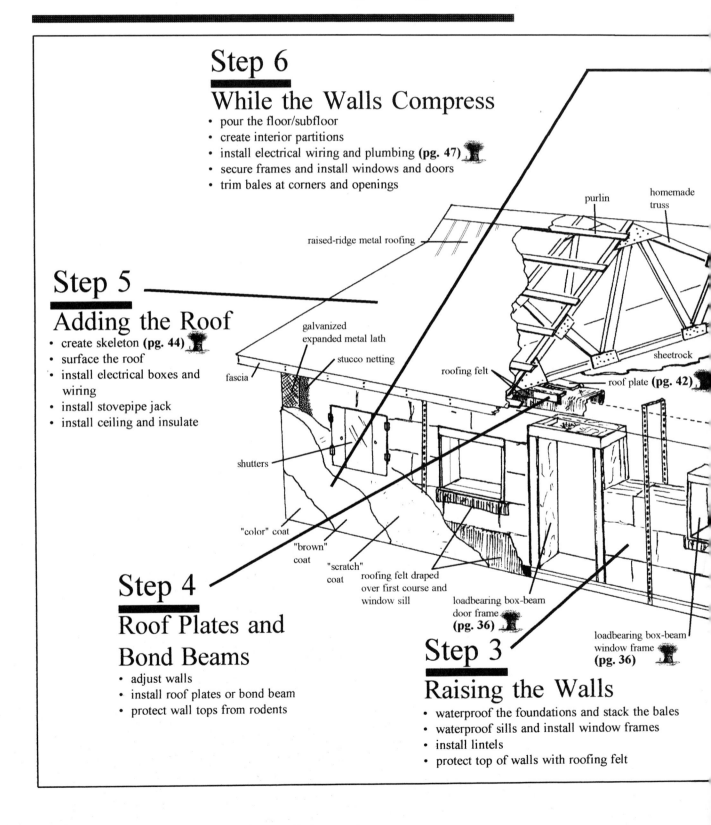

Step 6
While the Walls Compress
- pour the floor/subfloor
- create interior partitions
- install electrical wiring and plumbing **(pg. 47)**
- secure frames and install windows and doors
- trim bales at corners and openings

Step 5
Adding the Roof
- create skeleton **(pg. 44)**
- surface the roof
- install electrical boxes and wiring
- install stovepipe jack
- install ceiling and insulate

Step 4
Roof Plates and Bond Beams
- adjust walls
- install roof plates or bond beam
- protect wall tops from rodents

Step 3
Raising the Walls
- waterproof the foundations and stack the bales
- waterproof sills and install window frames
- install lintels
- protect top of walls with roofing felt

purlin

homemade truss

raised-ridge metal roofing

galvanized expanded metal lath

stucco netting

fascia

roofing felt

sheetrock

roof plate **(pg. 42)**

shutters

"color" coat

"brown" coat

"scratch" coat

roofing felt draped over first course and window sill

loadbearing box-beam door frame **(pg. 36)**

loadbearing box-beam window frame **(pg. 36)**

Step 7

Surfacing the Walls

- install stucco netting on outside
- install expanded metal lath inside and out
- stucco outside walls **(pg. 49)**
- plaster inside walls

Step 8

The Finishing Touches

- finish wiring and plumbing details
- finish carpentry details inside and out
- make the house a home

screened louvered attic vent

external roof plate tie-down

angle-iron lintel **(pg. 36)**

non-loadbearing door frame **(pg. 36)**

rebar stubs ("imbalers")

high-mass floor

pressure-treated wood nailer strip

reinforced concrete grade beam **(pg. 31)**

Step 1

Layout and Foundations

- establish corner locations
- erect batter boards and string lines
- install "rough" plumbing and electrical
- create the foundation for the bale walls with appropriate "hardware" in place

dry rubble footing **(pg. 31)**

Step 2

"building-grade" bale

Window and Door Frames

- fabricate all frames
- attach door frames to foundation

Step 1. Foundations

Challenge: to create a stable, durable base for the bales that will minimize the likelihood of water reaching them from below and of stress being put on the wall-surfacing materials by uneven settling or lifting.

Walk-Through

- *Orient building layout, giving consideration to passive/solar strategies and other concerns.*
- *Stake corner positions using 3-4-5 squaring technique or equivalent.*
- *Create batter board system, fine tuning for square and level.*
- *Mark ground with lime or equivalent to guide excavations.*
- *Remove strings and excavate.*
- *Fill excavations to the surface with uniform "rubble" material. Compact as needed.*
- *"Form up" to pour foundation, giving consideration to rebar placement, waterproof perimeter insulation board, stucco netting attachment, plumbing and other passageways.*
- *Mix, pour, settle and screed concrete, placing foundation bolts and rebar "imbaler" stubs before it becomes too stiff. If possible, place no foundation bolt (for threaded rod tie-downs) closer than 1-1/4 bale lengths from a corner so all corner bales can be placed and adjusted without hassle. For tie-downs closer than 1-1/4 bale lengths, use an external system.*
- *Trowel the area the bales sit on to a flat, relatively smooth surface, then keep moist for a maximum-strength cure. Any concrete that will remain exposed should be troweled to the desired finish at this time.*
- *After removing the formwork, modify the ground surface to assure good drainage away from the foundation.*

Dimensioning Your Foundation

As discussed earlier under *Finalizing Your Design*, most builders choose not to use foundation dimensions that are arbitrary or based on some non-bale-related module. They do this to avoid having to create many custom-length bales in each course and to avoid having these shortened bales break up the "running bond" (where each bale overlaps the two bales below it by nearly equal amounts). The preferred approach is to let the chosen bale layout for the first course and the "effective bale length" (see page 23) dictate the length and width measurements for whatever platform the bales will sit on. It's better to have this "foundation" slightly oversized, since stuffing loose "flakes" of straw into occasional small gaps is much easier than re-tying bales to shorten them. Based on the width of the actual bales you use, the width of any "grade beam" collar or "toe-up" platform will be about 18 in. for 2-tie bales and about 23 in. for 3-tie bales, including the width of any waterproof perimeter insulation.

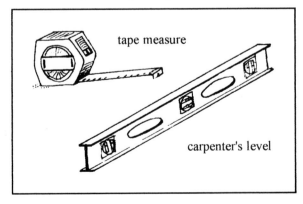

tape measure

carpenter's level

Site Layout

The purpose of layout is to accurately establish the location of the corners of the outside edge of the element (e.g., slab, grade beam, wooden deck) on which the bottom course of bales will rest.

The use of batter boards and string lines enables the builder to re-establish these corner points even though corner pins initially placed in the ground have been disturbed or removed. By positioning the horizontal cross-members of the batter boards at the same elevation (using, for example, a hose level and a carpenter's level), the strings can then also be used as a "bench mark" from which one can measure down to establish the correct depth of a trench or the correct height of formwork for containing poured concrete. Since small errors can be cumulative during the building process, it make sense to insure that the layout accurately reflects the dimensions and shape shown on your final drawings. However, most straw-bale builders feel comfortable with diagonals (corner-to-corner measurements) that differ by as much as a half inch.

References we recommend on site layout are Law (1982b), Wagner and Kicklighter (1986), and Jackson (1979).

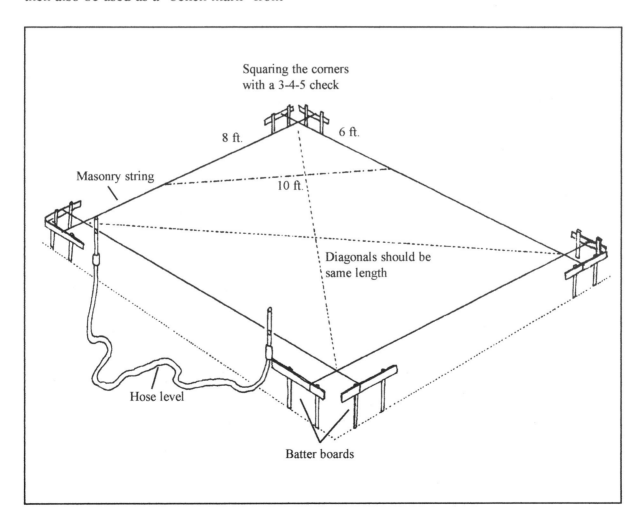

Forming Your Foundation

Wooden (pressure-treated preferred) nailer strip temporarily attached to the framework (and backed with spikes for firm connection to concrete) for fastening stucco netting and building felt. Other options shown below.

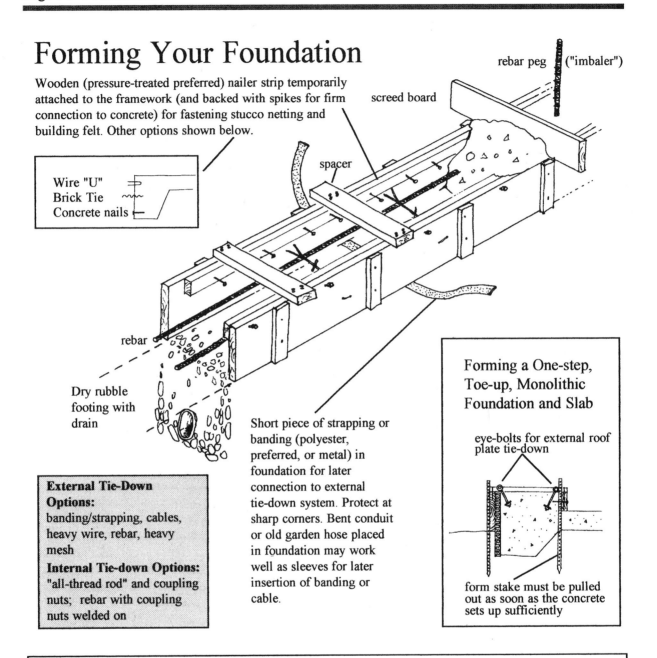

rebar peg ("imbaler")

screed board

spacer

Wire "U"
Brick Tie
Concrete nails

rebar

Dry rubble footing with drain

Short piece of strapping or banding (polyester, preferred, or metal) in foundation for later connection to external tie-down system. Protect at sharp corners. Bent conduit or old garden hose placed in foundation may work well as sleeves for later insertion of banding or cable.

External Tie-Down Options:
banding/strapping, cables, heavy wire, rebar, heavy mesh
Internal Tie-down Options:
"all-thread rod" and coupling nuts; rebar with coupling nuts welded on

Forming a One-step, Toe-up, Monolithic Foundation and Slab

eye-bolts for external roof plate tie-down

form stake must be pulled out as soon as the concrete sets up sufficiently

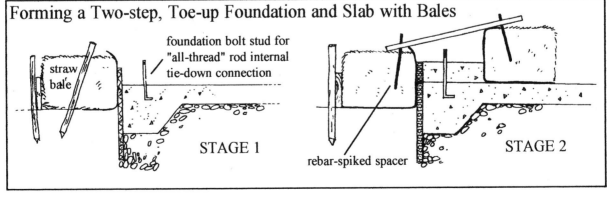

Forming a Two-step, Toe-up Foundation and Slab with Bales

straw bale

foundation bolt stud for "all-thread" rod internal tie-down connection

STAGE 1

rebar-spiked spacer

STAGE 2

Concrete

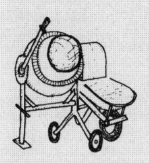

Concrete is a chemically-hardened mixture of cement, sand, gravel and water. A standard mixture is 1 part Portland cement, 2 parts sand, and 3 parts gravel.

Make sure that your forming system is level and strong enough to withstand the very considerable outward (and to a lesser extent upward) pressure that will be put on it by the wet concrete. Make sure that any passages through the concrete that will be required for later installation of pipes or electrical wires have been accounted for. Mark your formwork with some easily visible code system that shows where various items of hardware (*e.g.*, rebar stubs, eye-bolts, foundation bolts) need to be inserted into the still-wet concrete.

If using site-mixed concrete, consider equipment, labor and time requirements and local availability of acceptable sand and gravel. If using truck-delivered, already-mixed concrete, consider access for the truck and its chute, helpers and equipment needed to handle a large amount of concrete in a short time and where you might beneficially use any excess. For additional tips, consult Kern (1975), Syvanen (1983), MWPS (1989) or Loy (1990).

Calculate the cubic yards of concrete needed by multiplying the length by the width by the height of the foundation and/or slab (in feet), then dividing by 27 (the number of cubic feet in a cubic yard). Add 10% to the calculated amount to ensure having enough.

Foundation Strategies for Cold Climates

One downside of our inevitably wide bale walls is that any concrete platform on which they rest must also be wide. In areas where freezing temperatures are encountered at considerable depth, it would require large amounts of concrete to create a uniformly-wide concrete footing extending to below this "frost line". The related costs, both financial and environmental, dictate that we explore alternatives.

One possible solution is the "I-beam" concept, suggested to us by architect Arlen Raikes. The "I" cross-section, being narrower in the middle, requires the use of less cement-based materials.

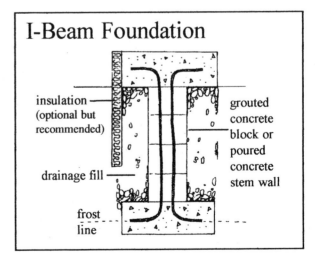

I-Beam Foundation

insulation (optional but recommended)

drainage fill

frost line

grouted concrete block or poured concrete stem wall

Another approach, suggested to Frank Lloyd Wright (1954) by Welsh-born masons in

Wisconsin, is the "dry wall footing". As Wright used this strategy, trenches were dug only sixteen inches deep to contain "rubble", even though the frost line depth was about four feet below ground surface. He assumed, apparently correctly, that if the material under the "grade beam" (or an equivalent collar) could be kept dry, there could be no destructive "frost heaving" even if the soil temperature dropped below freezing. Modern users have generally chosen, or been required by building officials, to dig the trenches to below the frost line depth. This requires the use of more "rubble" to fill the trenches to the

surface, but "rubble" is less costly to buy and place than concrete. A good reference on the modern use of this approach is Velonis (1983).

The word "rubble", as used in this connection, includes a variety of coarse, quickly-draining materials. Wright used "fist-sized broken rock" in his trenches. Modern builders have used everything from "river run" (rounded pebbles which settle automatically to a stable configuration), to "leach field rock" (angular fragments that may require mechanical compaction).

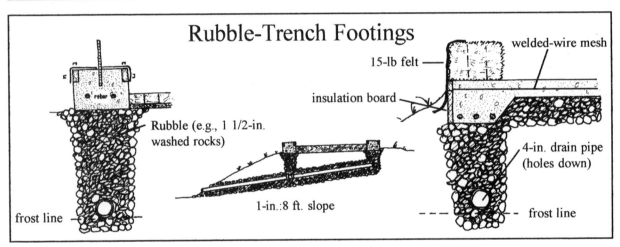

Rubble-Trench Footings

A third approach (shown below) is the "shallow, frost-protected footing" concept described in a publication prepared by the National Association of Home Builders Research Center for the US Dept. of Housing and Urban Development (HUD 1994). This approach is based on the use of

strategically placed waterproof, foam-board perimeter insulation to modify the frost line surface such that a frost-free zone is created under and around the edges of the building "footprint". This enables the builder to safely position the bottom of the footing well above the "normal" frost line.

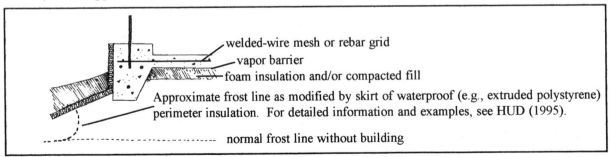

Other Foundation Options

Steeply-Sloped Lots

Sloping building sites present problems to the straw-bale builder. Using the "cut-and-fill" approach may require massive earth moving that results in ugly "cut" walls that needs retaining. Step footings have their own problems, including the possibility of differential settling caused by the differing number of bale courses in load-bearing structures. A better solution may be to use a grid of vertical columns or posts to support a wooden deck upon which the straw-bale house can sit. The space under the deck can be closed in with (straw-bale?) skirting, and used for storage (this would be a great place for storing water harvested from the roof). See Levin (1991) for further options.

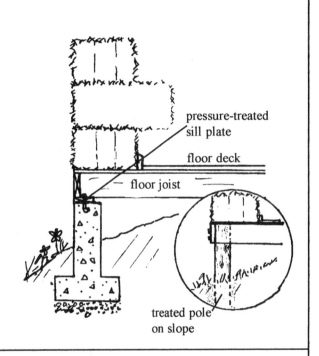

pressure-treated sill plate

floor deck

floor joist

treated pole on slope

Low-Tech Ideas

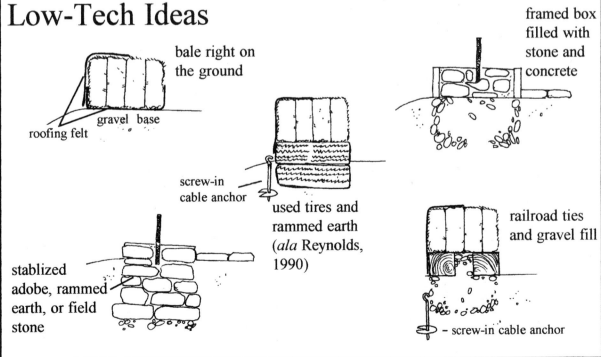

bale right on the ground

gravel base

roofing felt

screw-in cable anchor

used tires and rammed earth (*ala* Reynolds, 1990)

framed box filled with stone and concrete

stablized adobe, rammed earth, or field stone

railroad ties and gravel fill

– screw-in cable anchor

Step 2. Door and Window Frames

Challenge: to create a loadbearing frame, or non-loadbearing frame and loadbearing lintel, to accomodate each door or window in our straw-bale walls. In doing so, we must consider, at a minimum, the width of the specific opening, the nature of the weight above the opening, and the degree of compaction of our bales.

Walk-Through

- *Know, in advance, the actual sizes of all windows and doors.*
- *Fabricate the rough frames in advance of wall raising. Brace diagonally to keep them square until they are firmly secured in the walls.*

- *Position door and floor-mounted window frames onto the foundation. Once secured, provide temporary bracing to keep them upright and level.*
- *Fabricate any separate lintels that you will use above non-loadbearing frames.*

Loadbearing versus Non-Loadbearing Walls

Builders generally favor heavy-duty, loadbearing frames for openings wider than about 4-ft., especially in **loadbearing walls**.

They generally favor light, non-loadbearing frames with angle-iron lintels for openings less than about 4-ft. wide, especially in **non-loadbearing walls**.

Sizing Openings

1. Modify bales to fit arbitrarily positioned, framed, standard windows or pre-hung doors.

2. Make frames to fit openings dictated by the one-half bale module. Doors and windows will have to be custom made.

3. As in # 2, make frames to fit modular opening. Then, make a second, perhaps lighter, internal frame to fit a standard window or pre-hung door.

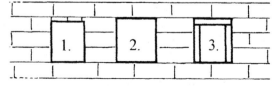

Window Seat Options

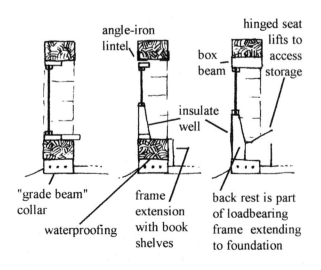

angle-iron lintel

hinged seat lifts to access storage

box beam

insulate well

"grade beam" collar

waterproofing

frame extension with book shelves

back rest is part of loadbearing frame extending to foundation

Loadbearing Options

Loadbearing, Box Beam Door Frame

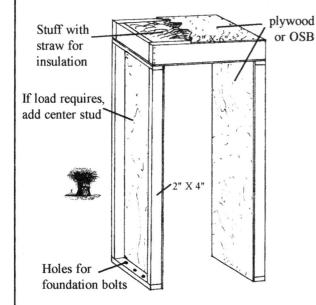

Stuff with straw for insulation

plywood or OSB

2" X 6"

If load requires, add center stud

2" X 4"

Holes for foundation bolts

Loadbearing Box Beam Window Frame

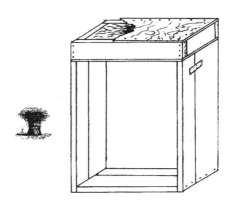

2" lumber. For wider openings (*i.e.*, greater loads) you may want to use box beam approach for sides and bottom assemblies, also.

Non-Loadbearing Options

Non-Loadbearing Door Frame and Lintel

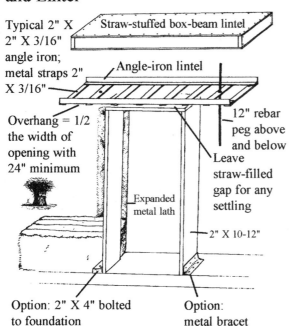

Typical 2" X 2" X 3/16" angle iron; metal straps 2" X 3/16"

Straw-stuffed box-beam lintel

Angle-iron lintel

Overhang = 1/2 the width of opening with 24" minimum

Expanded metal lath

12" rebar peg above and below

Leave straw-filled gap for any settling

2" X 10-12"

Option: 2" X 4" bolted to foundation

Option: metal bracet

Non-Loadbearing Window Frame

An additional bale-width piece of plywood over the opening may be advisable

Corner bracing

May need to re-tie if aggressively rounded or beveled

Galvanized expanded metal lath

Poured concrete interior sill

Sharpened 5/8" X 14"-long wooden dowel or 1/2" rebar pin

Roof Plate As Lintel

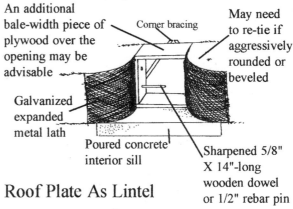

Rigid wooden roof plate (or concrete grade beam) acting as lintel

Angle-iron attached to both sides of standard roof plate

Step 3. Raising the Walls

Challenge: to create walls that match your expectations for function and form, in a way that reflects your interest in human interaction.

Walk-Through

• *Have all your building materials, hardware, and tools assembled on the site. If the weather permits, uncover your bale-storage stacks to give them a final chance to become fully dry.*

• *Seal the top of the foundation with a waterproofing "membrane" (e.g., roofing felt, plastic sheeting, various asphalt-based compounds, or combinations thereof). Be careful to seal around the protruding rebar stubs ("imbalers") and any foundation bolts. As desired—or as required—install a termite barrier (e.g., galvanized sheet metal, diatomaceous earth).*

• *Install temporary, sturdy corner guides, as desired. These help keep the corners vertical and, with string lines pulled between them, can help keep your walls (especially long ones) straight and vertical. Attach them securely and check them **often** for verticality ("plumb").*

• *If a large number of people will be assisting, it will help to break them up into working teams. These can consist of an experienced "wall captain" (for ongoing problem spotting and quality control) and three or four inexperienced members (for bale inspection, carrying, placing and pinning). It also helps to have several two-person crews set up to make half- and custom- length bales for the wall crews. Encourage people to trade jobs occasionally.*

• *If you have chosen to have your wall-raising be a "Y'all come!", community-building event, you will probably want to cordon off the work site from the socialization area.*

• *Before any bales are laid, it may be valuable to ask the workers to be mindful of job safety, to review with them the "rules of thumb" for laying bales (see page 40), and to inoculate them against the insidious "bale-laying frenzy".*

• *Start laying bales at corners and on both sides of door frames, lining them up accurately with the edge of the "foundation". If an internal roof-plate-tie-down system is being used, some bales will have to be lowered down over all-thread or rebar rods. With practice, careful measuring and marking of the "insertion spot" on the bottom of the bale, one can generally get the bale to the desired location on the first try.*

• *Before starting the second course, many builders drape a layer of 15-lb. roofing felt down over the outside of the first course of bales to protect them from possible water damage. The bales in the second course will overlap the bales below to form a "running bond".*

• *You can increase the stability at corners by driving in one or two "staples", bent from short lengths of rebar, where the corner-forming bales butt in each course. They can also be used in situations where additional connection between bales is desired (e.g., above a lintel).*

• The last few bales that will complete a section of wall should be put in place temporarily so that you can measure the size of the gap, if any, or any overlap. If the gap or overlap is very small, you may be able to find and substitute longer or shorter bales. Otherwise, fill the gap (make into two gaps if larger than about 4 inches) with a "flake" from a "bad" bale, or adjust for the overlap by shortening a bale.

• If corner guides are not being used, make diligent use of a carpenter's level (attached to the edge of a straight board) to maintain verticality at the corners, the only part of the walls impossible to mechanically "tweak" (i.e., bash, pound, or push into place) after they are finished. Since the upper parts of the walls will lengthen slightly as the bales compress under roof load, some builders try to slope the tops of the walls slightly inward at the corners to compensate. You may need to temporarily brace long, tall walls, especially in windy areas.

• The door frames will have been fastened to the foundation before any bales are laid, but window frames, except in the rare case where they will sit on the "foundation", cannot be set in place until the proper wall height is reached. After a waterproof "membrane" has been placed over the wall at the correct location along the length of the wall, the frame can be positioned within the wall width as previously determined. Many choose to maximize the interior sill-shelf and minimize potential water damage by mounting the windows essentially flush to the outside surface. An exception might be south-facing windows, in a design without roof overhangs, where small windows can be shaded from summer sun by placement to the inside, above a well-sloped exterior sill. Whatever placement is chosen, the frames should then be leveled, both horizontally and vertically, before being temporarily braced.

• If aggressively beveled/rounded bales are to be used to widen the interior wall opening at doors or windows, they should be customized and placed on either side of the frames as the wall goes up. Minor rounding can be done after the walls are up. An alternative is to make the frame wider than the door or window and use carpentry to create the bevel on the sides of the opening.

• Bale pinning normally takes place as the walls are being raised, often starting at the fourth course. At window locations, short pins can be driven into the bales beneath the frame, either before or after it is placed on the wall.

• Above non-loadbearing frames, some kind of lintel will need to be used to bridge across the opening and distribute the roof and/or wall weight from above the opening to the bale walls on either side. A generally accepted "rule of thumb" is that, at least in loadbearing walls, the lintels should extend out onto the walls on both sides for a distance equal to about half the width of the opening. Increase this distance if the bales at the opening are significantly rounded or beveled.

• Every few courses, check for level and shim with loose straw if necessary. Stuff gaps where bales butt with loose straw or scraps of fiberglass insulation. Do not force straw into gaps; it can force wall out of line.

• When the walls have been raised to the desired height, a waterproof "membrane" should be placed along their tops to protect them from rain or snow until the roof has been sheathed. Many builders choose to leave this "cap" in place permanently to protect the top of the walls against eventual roof leakage.

Customizing Bales

Since the bales are stacked in an overlapping "running bond", customized shorter bales will be needed where the walls butt against door and window frames. As shown below, first create tight new ties to contain the mini-bales. Then cut the original ties, at the knot, to get the maximum reusable length of twine or wire.

Re-Stringing Bale On Edge

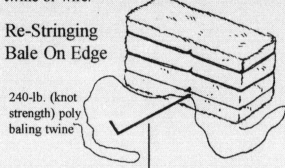

240-lb. (knot strength) poly baling twine

1/4" - 5/16" X 3' metal rod "needle tool", with one end pounded flat, drilled with 2 holes large enough for twine, then ground to a dull point

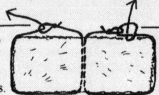

STEP 1.
Pull tight and clamp here with pliers or Vise-Grips.

STEP 2.
Snug tight, release clamp, then repeat knot.

Making Wedged Bales

1. Create new shorter ties to allow for bevel.
2. Add "keeper" strings that run to same string on underside.
3. Cut original strings at knot and remove.
4. Carefully remove unwanted straw. A chainsaw, bow saw, or hay saw works well for this.
5. The effect of beveling is to enhance penetration of sunlight and keep narrow openings from feeling like slots.

"keeper" string

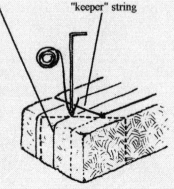

Building the Walls

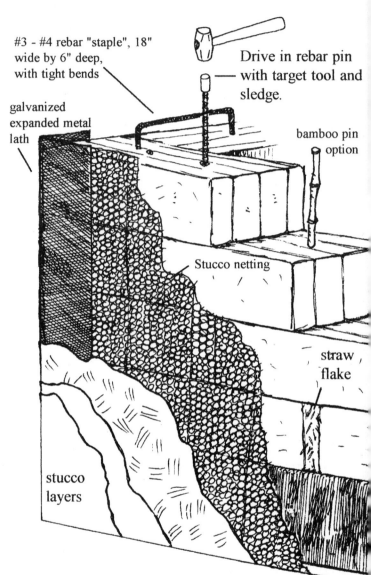

#3 - #4 rebar "staple", 18" wide by 6" deep, with tight bends

galvanized expanded metal lath

Drive in rebar pin with target tool and sledge.

bamboo pin option

Stucco netting

straw flake

stucco layers

Options for Target Tools

replaceable threaded cap

solid weld

1 1/4" by 3/16" plate

Square or round steel bar with hole. Lifetime guaranty!

gas pipe nipple

heavy-wall pipe (even then, cracks eventually at weld)

Rules of Thumb for Laying Bales

1. Start with good bales. Inspect each bale before placing it on the wall. Straighten if necessary. Use really "bad" bales for flakes.
2. On graph paper, map out bale and frame placement for each wall. Post these drawings prominently on the job site; consult often!
3. Start laying at corners and door frames; later at window frames.
4. Monitor corners carefully to keep them vertical (or sloping slightly inward).
5. Never cram a bale or a flake into place; it can move things out of plumb.
6. Take your time. Pay attention to details.

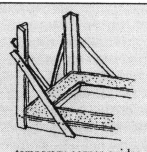

temporary corner guides

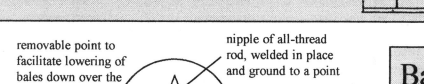

removable point to facilitate lowering of bales down over the threaded rod

nipple of all-thread rod, welded in place and ground to a point

coupling nut

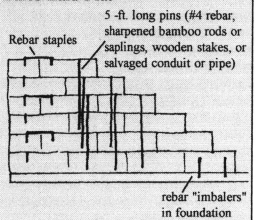

All-thread steel rod attached by coupling nuts to a foundation bolt and subsequent rod sections. This internal, roof plate tie-down system is popular, but expensive.

12" of exposed #4 rebar "imbaler"

asphalt roof cement

roofing felt

2" X 3 or 4" wooden strip fastened to top of grade beam to enable stapling of felt and stucco netting. Pressure-treated wood preferred.

Bale Pinning

Popular Pinning System for Loadbearing Designs with 3-Tie Bales Laid Flat

Rebar staples

5-ft. long pins (#4 rebar, sharpened bamboo rods or saplings, wooden stakes, or salvaged conduit or pipe)

rebar "imbalers" in foundation

Variations:

1. Short pins, two per bale, penetrating two courses at a time. Start with second course and repeat with all subsequent courses.

2. Eight-ft. long, sharpened "pins" of #5 rebar, two per bale, penetrating the full height of the wall. This technique provides no wall stiffening until the whole wall is up.

Step 4. Roof Plates and Bond Beams

Challenge: to create, for the top of each loadbearing wall, a roof plate or bond beam element that can be securely attached to the foundation and to which can be attached the roof structure. These elements should be sturdy enough to transfer the focused loads of each rafter, *viga* or truss to the wall below without unacceptable bending or cracking.

Walk-Through

• *Mechanically "tweak" (i.e., beat, bash, pry, brace) the walls until they are acceptably straight and vertical. Measure diagonally from corner to corner to check for square.*

• *Fabricate the roof plate on the ground in transportable sections (in some cases it is easier, and ultimately more accurate, to fabricate the plates directly on top of the foundation, beforehand).   Then move these sections to the top of the wall and connect them, taking care to get strong connections between sections (especially at corners).*

Make sure that the diagonals are also as nearly equal as possible and that the walls are properly aligned and secured under the plate. Then tie it down to the foundation, using whatever system was chosen.

• *Unless your roof plate / bond beam already adequately protects the top of the walls from invasion by rodents, deny them access by utilizing various materials (e.g., cement-based mortar, metal lath, sheet metal, plywood scraps, old boards) alone or in combination.*

Double-Layer Roof Plate

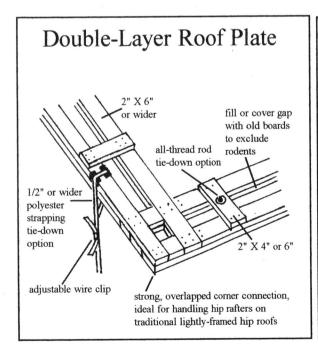

2" X 6" or wider

fill or cover gap with old boards to exclude rodents

all-thread rod tie-down option

1/2" or wider polyester strapping tie-down option

adjustable wire clip

2" X 4" or 6"

strong, overlapped corner connection, ideal for handling hip rafters on traditional lightly-framed hip roofs

Mongolian Prototype
(suggested by Dan Smith & Associates)

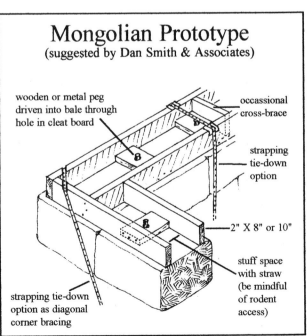

wooden or metal peg driven into bale through hole in cleat board

occassional cross-brace

strapping tie-down option

2" X 8" or 10"

stuff space with straw (be mindful of rodent access)

strapping tie-down option as diagonal corner bracing

Rigid Roof Plate

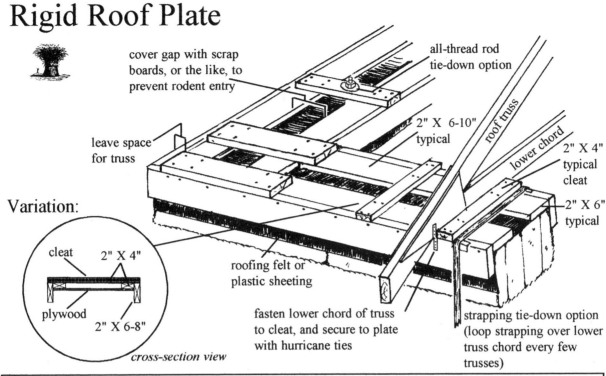

cover gap with scrap boards, or the like, to prevent rodent entry

all-thread rod tie-down option

leave space for truss

2" X 6-10" typical

roof truss

lower chord

2" X 4" typical cleat

2" X 6" typical

Variation:

cleat   2" X 4"

plywood

2" X 6-8"

cross-section view

roofing felt or plastic sheeting

fasten lower chord of truss to cleat, and secure to plate with hurricane ties

strapping tie-down option (loop strapping over lower truss chord every few trusses)

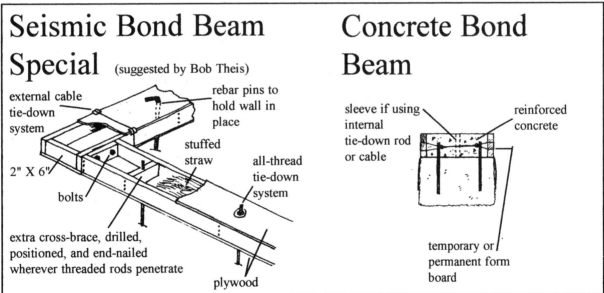

Seismic Bond Beam Special (suggested by Bob Theis)

external cable tie-down system

rebar pins to hold wall in place

stuffed straw

all-thread tie-down system

2" X 6"

bolts

extra cross-brace, drilled, positioned, and end-nailed wherever threaded rods penetrate

plywood

Concrete Bond Beam

sleeve if using internal tie-down rod or cable

reinforced concrete

temporary or permanent form board

Tie-Down Specifications and Placement

An engineer can calculate the total uplift force to be resisted and what the spacing along the wall should be for tie-down materials having certain strengths if such factors as roof area, roof overhangs, roof weight, maximum potential local wind velocities, etc., are known. Six-foot spacing for 1/2" threaded rod has often proved to be acceptable. Err on the side of caution!

If the roof plate is rigid enough, and the tie-down system sufficiently strong, the compression of the walls can be accomplished by pulling the roof plate/bond beam down, thus allowing immediate wall surfacing, if desired.

Step 5. Adding the Roof

Challenge: to create a sheltering cap (some combination of ceiling, and/or roof, and insulation) that is securely attached to the roof plate / bond beam. It needs to protect the tops of your walls and your interior spaces from the elements and keep heat from moving down into or up out of these spaces. For loadbearing designs, light roofs, bearing on all four walls, make real sense.

Walk-Through

- *Fabricate the central part of the roof skeleton, using identical homemade or commercial trusses. Complete the end hips, using hip trusses or traditional framing. Double up the two end trusses if your hip system concentrates a load on them.*

- *Brace the skeleton as it grows, leaving this bracing permanently in place where appropriate.*

- *Securely attach all trusses (and any rafters) to the outside edge of the roof plate using the appropriate connectors (aka hurricane ties or the equivalent).*

- *Attach 2" X 4" purlin strips, at 2-ft. intervals, to the roof skeleton.*

- *Fasten 26-gauge metal roofing to the purlins with special, self- tapping screws equipped with neoprene washers, using standard caulking strips where adjacent panels overlap.*

- *Create screened, louvered vents in the gablets at each end of the roof peak, installing proper flashing where the sloping metal roofing meets the vertical vent framing.*

- *Attach some material to the underside of the overhang created by the ends of the trusses/rafters, leaving adequate, screened vents to allow air movement up into the attic space.*

- *With the roof skin now in place, move inside and install any radiant heat barriers following manufacturer's directions.*

- *Install all necessary ducting, stove pipe brackets, electrical boxes (e.g., for overhead lights, smoke detectors, fans), wiring and plumbing in the attic space.*

- *Install the ceiling and insulate (or vice versa). In climates that require significant space heating or cooling, don't skimp here—equal or exceed the R-value of your walls, if possible. Be sure to include a reasonable access hole up into the attic space with a well-insulated cover.*

Photo by Phil Decker

Standard Roof Shapes

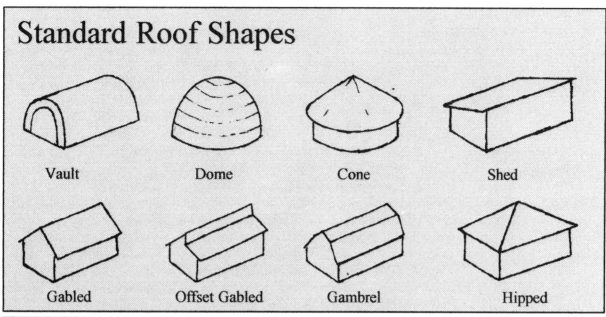

Vault Dome Cone Shed

Gabled Offset Gabled Gambrel Hipped

Dutch Hip Framing Options

This option is often used on rectangular buildings as a more interesting substitute for a simple gable roof. Depending on the framing system (three possiblities shown here), some roof weight can be distributed to the shorter end walls.

Another advantage is eave protection over all the walls.

Good written resources include Dunkley (1982); Law (1982a); Feirer and Hutchings (1986); and Gross (1994).

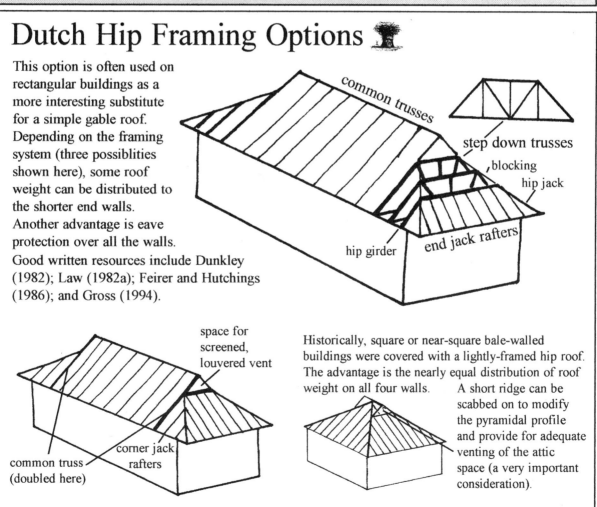

common trusses

step down trusses

blocking

hip jack

hip girder end jack rafters

space for screened, louvered vent

common truss (doubled here)

corner jack rafters

Historically, square or near-square bale-walled buildings were covered with a lightly-framed hip roof. The advantage is the nearly equal distribution of roof weight on all four walls. A short ridge can be scabbed on to modify the pyramidal profile and provide for adequate venting of the attic space (a very important consideration).

Some Truss Types

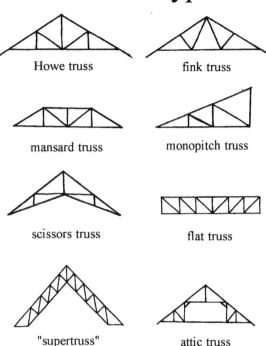

Howe truss

fink truss

mansard truss

monopitch truss

scissors truss

flat truss

"supertruss"

attic truss

For more, see Booth (1983), MWPS (1989a), or Smulski (1994).

Cathedral Ceilings

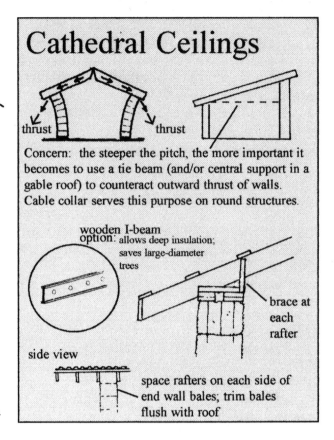

thrust   thrust

Concern: the steeper the pitch, the more important it becomes to use a tie beam (and/or central support in a gable roof) to counteract outward thrust of walls. Cable collar serves this purpose on round structures.

wooden I-beam
option: allows deep insulation;
saves large-diameter
trees

brace at each rafter

side view

space rafters on each side of end wall bales; trim bales flush with roof

Ceiling Insulation Options

Don't skimp here! Have enough to equal or, better yet, exceed the R-value of your walls (see Lenchek *et al.* 1987, and Nisson and Dutt 1985).

Options include the following:
- fiberglass
- foamboard
- foam-filled panels
- blown cellulose
- cotton fiber (batts or blown)
- air-krete
- loose straw, flakes or bales

If indicated by your climate, use an air barrier to prevent water vapor in the rooms from moving up into the insulation to condense and freeze, thus greatly reducing its effective R-value.

Roof Surface Options

Having a secure and durable roof surfacing is perhaps the single-most important factor for the long-term structural health of your straw-bale home. Roof surfacing options are numerous, including metal, roll roofing, asphalt shingles, and shakes (see Herbert 1989).

A long-held desire of many straw-bale aficionados has been to simplify the roof structure to where much less wood is used while retaining adequate insulation. Vaults and domes may work. Another idea, using ferro-cement and an elastomeric membrane, is shown below:

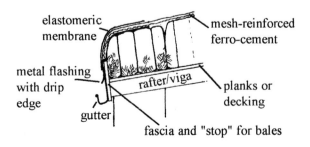

elastomeric membrane

mesh-reinforced ferro-cement

metal flashing with drip edge

rafter/viga

planks or decking

gutter

fascia and "stop" for bales

Step 6. Letting the Walls Compress

Challenge: to use this opportunity to work comfortably inside your building on a variety of tasks. Builders faced with a tight time schedule may consider using a tie-down system and a sufficiently rigid roof plate/bond beam that allows for rapid, mechanical compression.

Walk-Through

• *Unless your design includes a rigid roof plate / bond beam and a tie-down system that enables mechanical tightening, you must now let the walls undergo gradual compression in response to the "dead load" of the roof/ceiling system. The initial response is rapid, but then begins to taper off. Recent experience suggests that anywhere from three to about ten weeks may be needed for the walls to reach equilibrium. Depending on bale compaction and roof weight, total compression will vary from some fraction of an inch to several inches. During the settling period, you should occasionally adjust your tie-downs to remove any slack.*

• *With the tops of the bale walls now protected by the new roof and the bottoms sitting safely up off the ground on a waterproofed foundation and draped with plastic sheeting or roofing felt, you can catch your breath. Use this respite for things like:*

+ *recreating and reconnecting with loved ones.*
+ *tweaking any ornery bales into final position.*
+ *adjusting the verticality of door and window frames, as needed, and connecting them securely to the bale walls with dowels or metal pins.*
+ *installing the doors and windows.*
+ *creating the high-mass, finished floor, at an elevation 3 to 4 inches below the top of the grade-beam collar upon which the bales are sitting.*
+ *creating non-loadbearing interior partitions,*

leaving a gap above them to allow for the settling that may still take place. To maximize the ease of rearranging the position of interior partitions at some later time, some builders postpone this task and any electrical or plumbing related to partition walls until after the interior surfacing is in place on the exterior walls.

+ *extending the plumbing into the interior space, preferably in 2"x6" frame walls shared by two bathrooms or a bathroom and a kitchen.*
+ *equipping the straw-bale walls with wooden elements to enable hanging cabinets, bookcases, etc. If these elements will be hidden by the stucco/plaster, map their position precisely on a diagram and save it for later use.*
+ *installing electrical boxes and wiring, at least in the exterior walls, as called for on the electrical plan.*
+ *rounding/trimming off bales at exterior corners and at door and window openings, as desired, to provide for a "soft-profile" finished appearance. This is also the time to trim off any undesired protrusions on your wall surfaces (a line trimmer/"weed whacker" works beautifully for this). Any niches, notches, alcoves, etc., should all be created at this time using a small chain saw (or any effective substitute).*

Bale Tweaking Tools

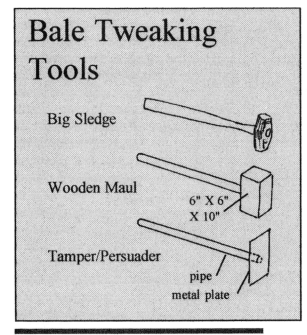

Big Sledge

Wooden Maul

6" X 6" X 10"

Tamper/Persuader

pipe

metal plate

Mounting Electrical Boxes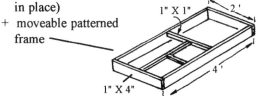

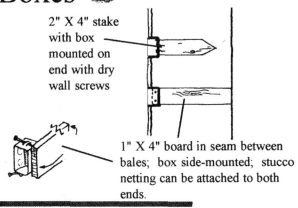

2" X 4" stake with box mounted on end with dry wall screws

1" X 4" board in seam between bales; box side-mounted; stucco netting can be attached to both ends.

Wiring Options

• Wires in metal or flexible plastic conduit on surface or in groove (dug out with claw hammer, tip of chainsaw, circular saw).
• Plastic-sheathed cable (*e.g.*, Romex 12/2) pushed into 2 1/4"-deep groove cut into walls. Hold in place with "Robert pins" (tightly bent, heavy wire U's).
• Cable run horizontally on a bale course during wall raising. Position about 3" from inner edge (CAUTION: be careful not to hit wire when pinning bales).

High-mass Floor Options

• Brick-on-sand (see Ring 1990)
• Tiles on slab
• Poured adobe (see Southwick 1981; Steen *et al.* 1994)
• Compacted soil cement/rammed earth (see Berglund 1985, McHenry 1989)
• Concrete
  + regular slab
  + scored or embossed slab (pressed-in pattern)
  + large, thick poured-in-place "tiles" (frame stays in place)
  + moveable patterned frame

1" X 1"   2'

4'

1" X 4"

Coloring Concrete

• Mix dye with the concrete before pouring.
• Sprinkle on and "float" in during final finishing.
• Staining: commercial or homemade (artist pigments or use ferrous sulphate which is cheap and available from agricultural or fertilizer suppliers to get a yellowish, reddish brown).
• Paint with special concrete paints.

Options for Interior Partitions

• Standard 2" X 4" or 2" X 6" frame with sheetrock or paneling (dowel to bale wall).
• Widely spaced frame, or shipping pallets, with light clay/straw infill.
• Wattle (bamboo/reeds/saplings) and daub (mud and straw).
• Hanging dividers.
• Furniture walls (bookcases, freestanding closets, etc.)
• Adobe (see McHenry 1989)
• Half-width straw bales

Step 7. Surfacing the Walls

Challenge: to provide long-term protection, both inside and out, from the elements, the occupants, infestation by vermin and depredation by curious cattle or vandals. If you plan to use wire-reinforced stucco or plaster, build adequate time into your schedule for the labor-intensive process of attaching the netting to the walls.

Walk-Through

• *With the settling nearly complete, this is a good time to insert any plastic pipe sleeves (for utility entrances) through gaps where the bale ends butt and to check to ensure that all other gaps have been stuffed with some insulating material. Then make a final inspection of your wall surfaces. Do any final trimming and filling of gaps or depressions. A mixture of chopped straw and mud makes a cheap filler that bonds well to the bales. (If you have chosen not to use wire netting on your walls, you may wish to let the stucco or mud plaster extend slightly into the gaps where adjacent bales butt to better "key" the surfacing material into the wall).*

• *You can now begin the process of surfacing the walls by creating an exterior curtain of "stucco netting" securely attached at the top to the roof plate and at the bottom to the wooden "nailer" in the side of the grade-beam collar. Where the horizontal strips of netting overlap, fasten them together with twists of galvanized wire or "cage clips" (aka "hog wire rings", short pieces of galvanized wire bent into a "C" shape; available at feed stores).*

• *Secure this curtain to the bale wall with galvanized wire (or an equivalent) pushed through the wall with a homemade "needle*

tool", once or twice per bale being typical. Springy areas between these through-ties can be snugged to the bales with long, narrow wire staples (often called "Robert pins" to distinguish them psychologically from "bobby-pins").

• *Wrap the building's corners with galvanized, expanded metal lath (aka "diamond lath"), to reinforce the stucco at these often-bumped locations. BEWARE: the cut edges of this stuff are like many, tiny razor blades.*

• *Cover all exposed metal and wood that will be covered by stucco with roofing felt or some other water-proof material. Then cover this material with galvanized, expanded metal lath, attaching it well to any wood backing or frames and to the stucco netting. Some builders are adding a commercially available metal edging strip, called "J-strip", to provide a uniform way to end stucco or plaster against frames.*

• *Design in one or more "truth windows" on interior (and/or exterior) to provide irrefutable evidence that your building is made of bales. Glass or plexi-glass in a frame works well, as do small, salvaged windows. Consider covering interior "truth windows" with art work in a side-hinged frame.*

• *Cement-based, exterior stucco can now be hand-applied, or blown on by a pumper rig (which is, unfortunately, not generally available for rent to ordinary folks). Typically, three coats are applied, the final, thin coat being dyed with a pigment. To achieve maximum strength, keep each coat of fresh stucco moist until full cured (about 48 hours). Although the "model" building being tracked here has mud plaster only on the interior walls, both stabilized (i.e., water-resistant) and natural mud plasters have been successfully used on the exterior walls of straw-bale walls (the latter needs ample roof overhangs in wetter climates).*

• *Since the plans for the building call for un-stabilized mud plaster applied directly to the straw without reinforcement on the interior surfaces of the straw-bale walls, you need only first cover with roofing felt and expanded metal lath any metal or wood that will be covered by the plaster, using the techniques employed on the exterior.*

• *The mud (aka adobe) plaster can now be applied directly onto the bales with a trowel or your hand, taking care to press the mud firmly into all depressions, cracks and crannies. Typically, two or three layers are applied, the last often being a clay slip that provides a smooth, uniformly colored surface.*

A Jungle of Jargon

• **Stucco** – usually refers to a hard wall-surface coating material applied as a mixture of portland cement, sand, lime and water. Can be used on both exterior and interior surfaces. Can be dyed, stained, painted and sealed. The term is sometimes also used for finish coat mixes that are commercially available.

• **Plaster** – a softer interior or exterior wall-surface coating often consisting of lime, sand and water (aka lime plaster). Gypsum plaster, a mixture of gypsum, lime and water, is for interior surfaces only.

• **Mud Plaster** – (aka adobe plaster) consists of a appropriate soil mixture (sand, silt and clay) mixed with water. Used traditionally in arid and semi-arid regions on both interior and exterior surfaces.

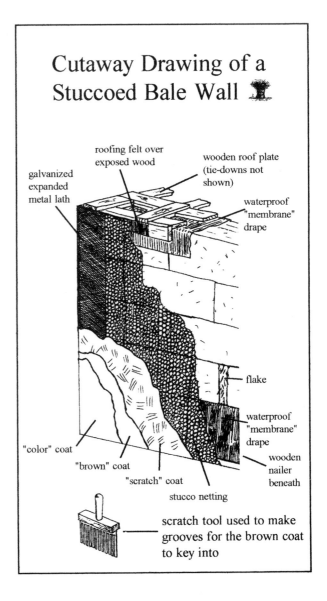

Cutaway Drawing of a Stuccoed Bale Wall

roofing felt over exposed wood

wooden roof plate (tie-downs not shown)

galvanized expanded metal lath

waterproof "membrane" drape

flake

"color" coat

"brown" coat

"scratch" coat

stucco netting

waterproof "membrane" drape

wooden nailer beneath

scratch tool used to make grooves for the brown coat to key into

Flow Chart for Wall-Surfacing Decision-Making

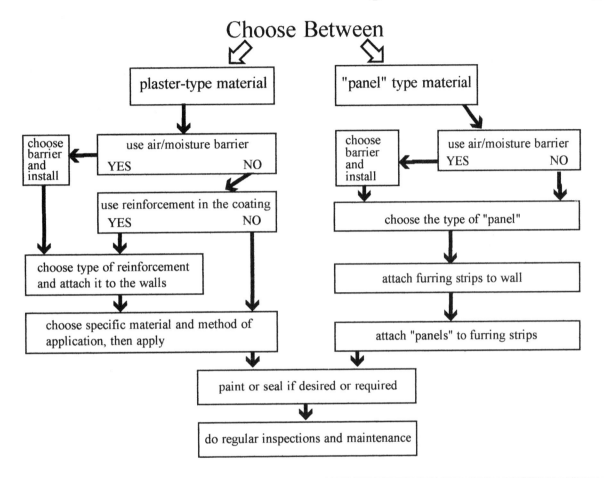

Air and Moisture Barriers

Purpose – to prevent the passage of air (and water vapor) and/or liquid water into the bale walls. Water vapor can eventually condense if the temperature in the walls drops low enough.

Common Types – plastic sheeting (air and water), vapor barrier paints (air and water), breathable house wraps (wind and water).

Possible Placement – Air barriers are often placed only on the interior surface, although occasionally on the outside as well. They may dictate the use of an air exchanger. Breathable house wraps are generally used only on the outside surface.

Pros and Cons – Unnecessary and not recommended in temperate climates, since they prevent the stucco or plaster from "keying" into the roughness of the bales and keep the walls from "breathing". In climates with simultaneous wind and rain or drifting snow, a house wrap extending part way up the wall should be considered. In colder climates, air barriers may be needed on the interior surfaces. See Gibson (1994) and Lstiburek and Carmody (1993) for more detail.

Reinforcement for Plaster/Stucco

Purpose - to help hold the "scratch" coat in place, to reduce cracking, to help tie down the roof plate, to sandwich the walls for increased resistance to dislocation by seismic forces.

Common Types - 1" chicken netting, stucco netting (heavier wire than in chicken netting), galvanized expanded metal lath (aka diamond lath). Plastic mesh, normally used to reinforce synthetic stucco coating sprayed onto siding or foamboard (*e.g.*, the Drivit System), may provide an alternative for those reluctant to use metal netting.

Pros and Cons - We strongly encourage the use of expanded metal lath: 1) where stucco/plaster butts up against or covers roofing felt, metal or wood; 2) on outside and inside corners of door and window openings; 3) on all outside wall corners.

Covering both surfaces of the bale walls has costs in time, labor, money and resources. You will have to decide, in your particular situation, seismic and otherwise, just how much "insurance" you want. Both cement-based stucco and mud plaster have been used successfully on bale walls without the benefit of reinforcement; however, the track record is still minimal.

Attaching the Reinforcement

Cage Clip held for placement and crimping by special pliers (you can modify your own) where strips of netting overlap or expanded metal lath overlaps wire netting

"Pine City Male Hooker" - push in parallel to flakes, turn 90% (as shown), pull back until hook engages, cut off just outside netting, bend end over a netting wire

"Robert Pins"

tightly bent, heavy coat hanger wire or the equivalent (10-12 ga., 6"-10" long)

Modified Screwdriver - where flaps of expanded metal lath extend out onto stucco netting, attach the edge of the flap by cutting small slits into its edge and using this tool to bend the resulting corners in around netting wires

Through-The-Wall Ties

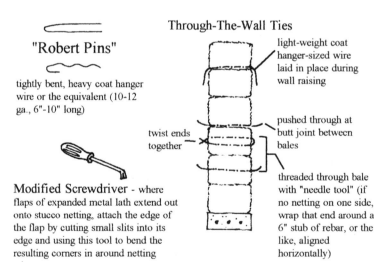

light-weight coat hanger-sized wire laid in place during wall raising

pushed through at butt joint between bales

twist ends together

threaded through bale with "needle tool" (if no netting on one side, wrap that end around a 6" stub of rebar, or the like, aligned horizontally)

Expanded Metal Lath Meets Wood Frame

lath and plaster

If exterior, include waterproof membrane between stucco and wood

wood frame

caulk

metal J-strip

1" roofing nails work well

* For these options attach the expanded metal lath before placing the frames

Recipe for a Cement-based Stucco

A common recipe for the "scratch" and "brown" coats of the traditional Southwest 3-coat finish calls for:

- 1 part (by volume) Portland cement
- 6 parts washed, clean, plaster sand
- 1 part lime
- chopped fiberglass threads, an optional additive intended to reduce cracking

Nowadays, the third, "color" coat is usually a store-bought mix with additives to provide a smooth colored finish.

Recipe for a Lime-based Plaster

Before cement was widely available, softer, lime-based plasters were commonly used on sun-dried adobes. Recent experience with them in France has been positive (soft, more breathable plaster on soft, breathable bales). As a start, for experimentation, try:

- 3 to 5 parts (by volume) sand
- 1 part hydrated lime

Stabilized Mud Plaster

Mud plaster may do poorly on exterior applications unless stabilized (*i.e.*, made water-repellent) or protected by generous roof overhangs or porches. The most common method of stabilizing mud plaster is to incorporate a pre-determined amount of a water-based asphalt emulsion into the water that is being mixed with the soil. This asphalt emulsion is a specialized industrial product not usually available except at bulk distributors of petroleum products (*e.g.*, CCS-1 from Chevron) or for companies specializing in road paving.

A process for determining the correct amount of emulsion to use with a given amount of any particular soil mixture is described in Tibbets (1989). Using too much will increase your cost and weaken the plaster. Using too little provides less than optimum water resistance.

Another approach is to try a standard formula or procedure. One adapted from Jeff Smith (Tibbets 1989), is as follows:

- Develop a "proper" mix of clay-rich soil and sand by experimentation (start at 1 soil to 4 sand sifted through 1/8" screen). Strive for a hard surface with few cracks.

- Experimenting in a mixing tub or wheelbarrow, use 7 full shovels of dry mix. Determine the amount of water needed to get the desired consistency, learned from playing around with your mud on the bales.

- Multiply this by 10.

- To a plaster mixer (a real labor-saver that can often be rented) add 3/4 of the amount of water previously determined, then add 2 1/2 gallons of emulsified asphalt.

- Now add 70 shovels of your mix or the correct number of shovelfuls to maintain the proper ratio of sand to soil that add up to 70.

- Add water in small amounts while mixing to finally reach the desired consistency.

- If possible, test the final mix on a small building or privacy wall first. Test for erodability with a pistol-grip sprayer on a hose. Use a patch of unstabilized plaster as a comparison.

Mixing, Application, and Curing

Manual. Small batches of your chosen stucco or plaster material can be mixed by hand in a mixing trough or wheelbarrow. A mortar hoe speeds up the process.

Stucco and plaster can be applied by hand using a hawk and trowel (see diagram, right). Mud plaster is sometimes applied directly with the palms of the hands.

Mechanical. A plaster mixer or even a standard concrete mixer can be a great labor-saver. Commercial plasterers often use a mixer/pumper to "blow" on these materials. This technique is fast and maximizes the penetration of the scratch coat into the roughness of the bales. If you can't afford having professionals doing the entire job, look into having at least the first coat blown on.

Coverage. As a very rough estimate, the first scratch coat of cement-based stucco (see page 52) directly onto a friend's bale walls required about 1 bag of cement per 43 sq. ft. of coverage; 1 bag of lime per 100 sq. ft.; and, about 5 yards of sand per 1000 sq. ft. of coverage.

Curing. Mud and lime plaster can simply be allowed to dry. Cement-based surfacing materials should be kept moist for 48 hours to ensure maximum strength.

Resources. Local stucco and plastering companies can often provide good advice about what works best in a given climate. For written resources, try Gorman (1988) and Pegg and Stagg (1976). For mud plasters, see McHenry (1989), Tibbets (1989), and Steen *et al.* (1994).

Gasoline-powered Plaster Mixer

Cutting From a Hawk

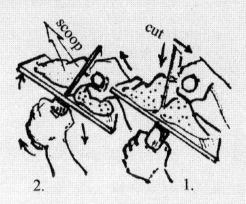

2.   1.

"It's tricky. You'll drop a lot before you learn. The secret is a certain twist of the hand and wrist, while tilting the hawk with a little motion...The motion of plastering is more like a sweep or arc, while using a pressure...Keep the hawk about one foot from the wall. It will be hard to keep plaster from falling off the hawk at first. Practice makes perfect!" *(from Tibbets 1989:63)*

Some Tools of the Trade

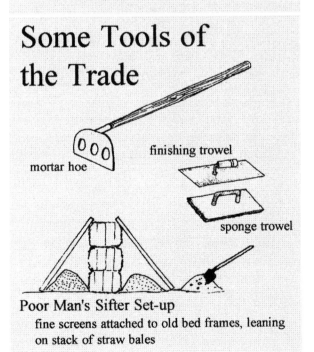

mortar hoe

finishing trowel

sponge trowel

Poor Man's Sifter Set-up
fine screens attached to old bed frames, leaning on stack of straw bales

Step 8. Finishing Touches

Challenge: to create interior and exterior environments that are low-maintenance, low-cost, flexible, practical, healthy, comfortable, visually pleasing, personal and nurturing.

Walk-Through

• Go back to Step 1, and this time, find an even simpler way to do it - a way that's customized to YOU as a builder/inhabiter. You'll have the advantage of knowing yourself better than any "guide-writer" could and of now being far out along the straw-bale learning curve. Really, don't you wish we'd started you off with a little storage shed or a stand-alone guest bedroom, rather than this big old thing. But, you've invested so much time, money, sweat and brain cells into getting to this point that, even though it's not perfect, you might as well finish it.

• Unless you chose to put interior partitions in before surfacing the interior walls, what you have now is an open space to be divided up according to your original floor plan (or some modification thereof). Consider doing the dividers (e.g., walls, screens) in such a way that they can be easily moved at some later time when your spatial needs and/or preferences have changed. Do any appropriate plumbing or electrical work before creating the finished surface on the partition walls.

• Most historic and modern bale builders have used thin frame walls, sheathed with gypsum plaster or sheet rock, to divide up their structures. They take up little of the interior floor space, and are cheap, quickly built, and easily moved or removed. If you go this route, consider filling the voids with something that will increase their thermal mass and/or reduce the transmission of sound through them (e.g., sand, tamped straw, tamped straw coated with a clay slip).

• If the natural color of the interior mud plaster is too dark for certain spaces, try a technique that author Steve came up with for his straw-bale home. Mix powdered dry-wall joint compound with water to reach a consistency just thin enough to paint on with a wide, stiff brush. Be sure to get the mud completely covered. When completely dry, roll or spray on an interior latex paint.

• Hang any wall-mounted cabinets, bookcases, etc., using the diagram you made earlier to locate the wooden elements they will attach to. Drive through (mud) or cut into the plaster to drive in dowels for the hanging of heavy artwork.

• Install any floor-standing cabinets, vanities, etc. Because of the inevitable irregularity of stucco/plaster on bales, some builders prefer to put all floor-standing units in place after the first (aka "scratch") coat is applied, making sure to first patch any cracks in the area to be covered by the units. The second (aka "brown") coat can then be used to fill the gap between the back edge of the counter, or splash board, and the wall *surface. Protect the units carefully with drop clothes.*

• Install plumbing-related items and associated fixtures, vents over air ducts, vent fans, lights/fans, wood stoves for backup heating, cooling devices, etc. Continue with the seemingly interminable installation of shelves, clothes rods and hooks, the sanding, puttying, sanding, caulking, priming, sanding, painting, staining, etc.

• Don't ignore the outside. Get some herbs and a kitchen garden in. Don't miss the right season to plant landscaping that will give you privacy, beauty, shade, and food (see Moffat et al. 1993, and Groesbeck and Striefel 1994).

Get your trees off to an early start. Add on any shade porches.

• Accept the inevitable truth that Step 8 actually never ends--it just continues until you realize that what you are doing would be better called maintenance. Now you are either ready to avail yourself of the 12-step program that Straw-bale Builders Anonymous offers, or to continue with our 8-step program on a new and more elegantly-simple, straw-bale project. Don't be surprised to see one or both of us at the SBBA meetings or lurking around your job site hoping to pick up some tips for the next version of this guide.

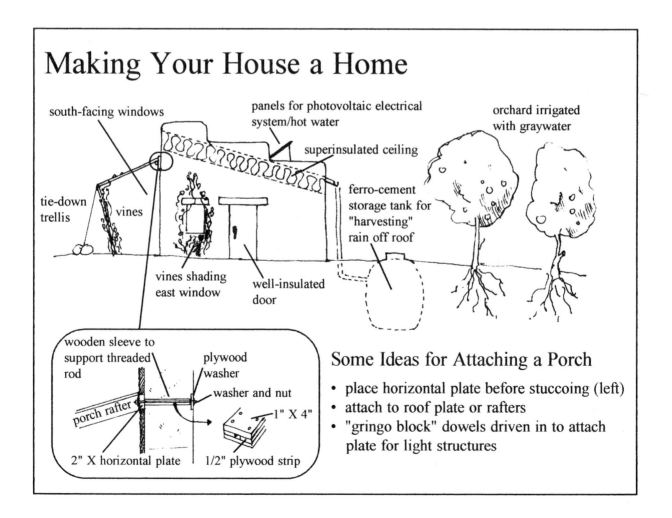

Making Your House a Home

- south-facing windows
- panels for photovoltaic electrical system/hot water
- orchard irrigated with graywater
- superinsulated ceiling
- tie-down trellis
- vines
- ferro-cement storage tank for "harvesting" rain off roof
- vines shading east window
- well-insulated door

wooden sleeve to support threaded rod
plywood washer
washer and nut
porch rafter
1" X 4"
2" X horizontal plate
1/2" plywood strip

Some Ideas for Attaching a Porch

- place horizontal plate before stuccoing (left)
- attach to roof plate or rafters
- "gringo block" dowels driven in to attach plate for light structures

The Non-Loadbearing Option

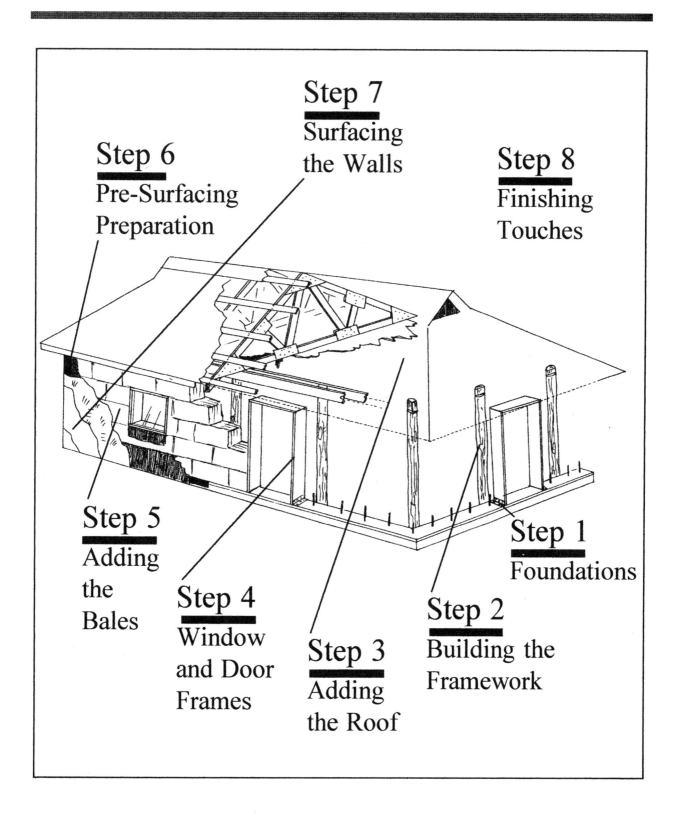

Step 7
Surfacing
the Walls

Step 6
Pre-Surfacing
Preparation

Step 8
Finishing
Touches

Step 5
Adding
the
Bales

Step 4
Window
and Door
Frames

Step 3
Adding
the Roof

Step 2
Building the
Framework

Step 1
Foundations

Step 1. Foundations

Challenge: basically the same as in a design with loadbearing walls. The details differ, however, since the roof weight is now transmitted to the foundations by some kind of framework. If the framework involves widely spaced vertical posts, the foundations must be designed to handle the concentrated loads transferred at these points. The foundation must also properly elevate and carry the bale walls.

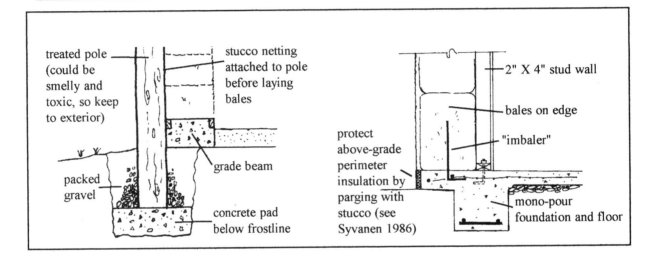

Step 2. Building the Framework

Challenge: to create a rigid, loadbearing framework to carry the roof weight and transfer it to the foundation. It should safely resist any horizontal (aka lateral) loads from wind or earthquakes. Multi-story structures become easily possible.

Possibilities for Loadbearing Frameworks

The possibilities run the gamut, from structural bamboo (a grass like straw), to traditional wooden frames (studs, timbers, poles, etc.), to concrete block columns with a poured concrete bond beam, to steel posts topped with glue-laminated beams, to thin masonry walls or panels. Most of these techniques are widely used and information on their "how to" is readily available. Recommended resources include Sherwood and Stroh (1992) and Wahlfeldt (1988) for wood frame; Benson (1990) for timber frame; NRAES (1984), Kern (1981), and Wolfe (1993) for post and pole.

Step 3. Adding the Roof

Challenge: basically the same as in the loadbearing option, although the use of a non-compressible framework does release you from some of the floor plan and roof weight constraints imposed when the walls are loadbearing.

Using Structural Roof Panels

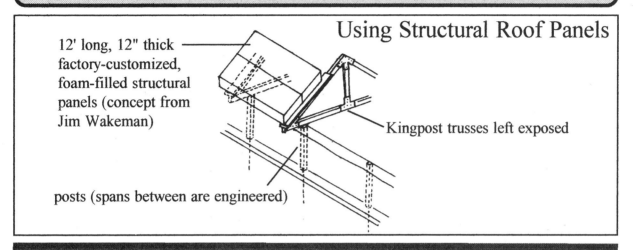

12' long, 12" thick factory-customized, foam-filled structural panels (concept from Jim Wakeman)

Kingpost trusses left exposed

posts (spans between are engineered)

Step 4. Window and Door Frames

Challenge: the same as for a loadbearing design. However, since no portion of the roof load is carried by the frames in a non-loadbearing design, and since all the needed wall rigidity can be built into the framework, there is freedom to make the openings larger and/or more numerous. The perceived desirability of this must be balanced against the relatively low R-value of doors and windows (even in the most expensive, high-tech models), and their effects on the performance of your superinsulated building.

Homemade, Double-Pane, Fixed-Glass Windows

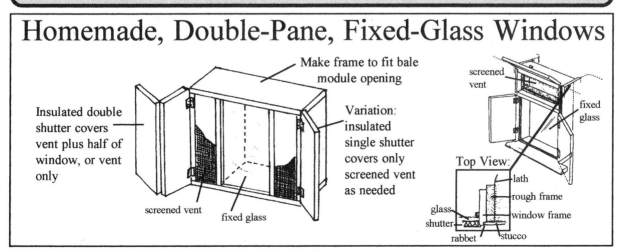

Make frame to fit bale module opening

Insulated double shutter covers vent plus half of window, or vent only

Variation: insulated single shutter covers only screened vent as needed

screened vent

fixed glass

screened vent

fixed glass

Top View:

lath

rough frame

window frame

glass

shutter

stucco

rabbet

Step 5. Adding the Bales

Challenge: similar to that with loadbearing walls, but the job of maintaining verticality as one stacks the bales is made much easier by the presence of the framework. On the other hand, you have the added task of fastening the walls to the framework. For certain designs, the bales must be notched to receive the posts.

Tools for Cutting Notches

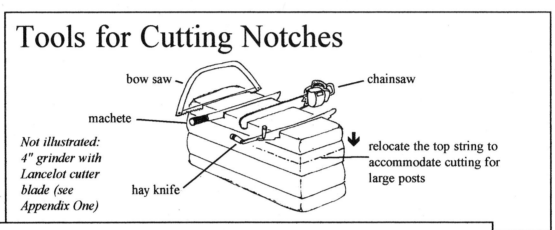

bow saw

chainsaw

machete

Not illustrated: 4" grinder with Lancelot cutter blade (see Appendix One)

hay knife

relocate the top string to accommodate cutting for large posts

Attaching to the Framework

Many approaches have been used. Use your creativity.

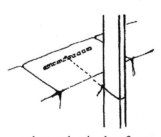

rebar stub wired to frame

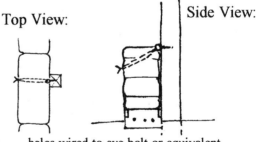

Top View:

Side View:

bales wired to eye bolt or equivalent
(technique especially helpful at corners)

Pinning the Bales

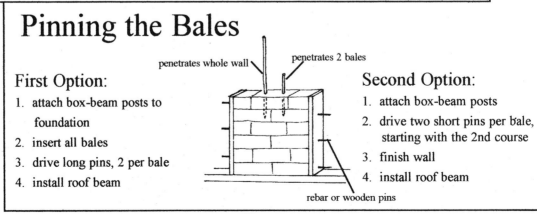

penetrates whole wall

penetrates 2 bales

First Option:

1. attach box-beam posts to foundation
2. insert all bales
3. drive long pins, 2 per bale
4. install roof beam

Second Option:

1. attach box-beam posts
2. drive two short pins per bale, starting with the 2nd course
3. finish wall
4. install roof beam

rebar or wooden pins

Options for Bale and Post Locations

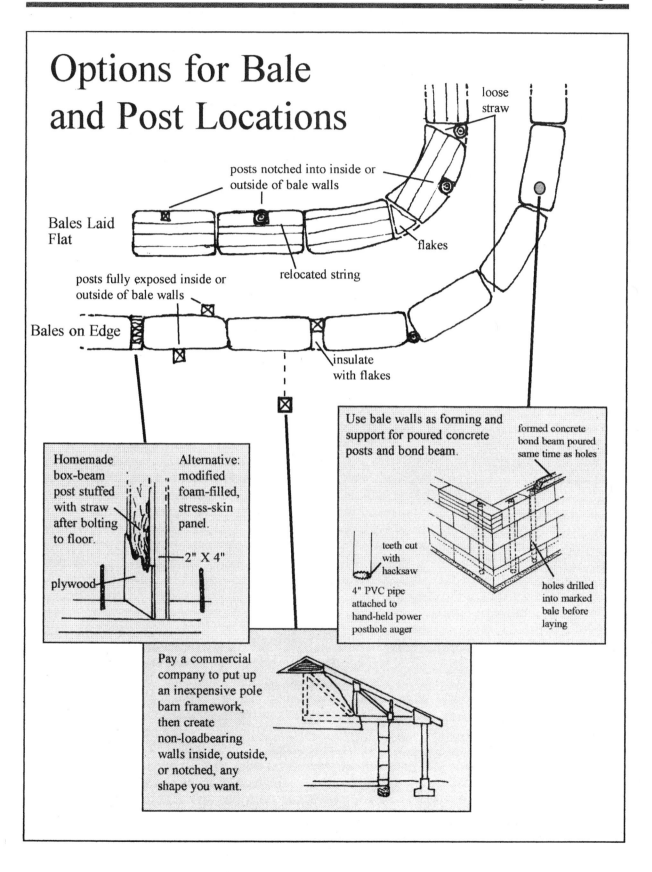

loose straw

posts notched into inside or outside of bale walls

Bales Laid Flat

flakes

relocated string

posts fully exposed inside or outside of bale walls

Bales on Edge

insulate with flakes

Homemade box-beam post stuffed with straw after bolting to floor.

Alternative: modified foam-filled, stress-skin panel.

2" X 4"

plywood

Use bale walls as forming and support for poured concrete posts and bond beam.

formed concrete bond beam poured same time as holes

teeth cut with hacksaw

4" PVC pipe attached to hand-held power posthole auger

holes drilled into marked bale before laying

Pay a commercial company to put up an inexpensive pole barn framework, then create non-loadbearing walls inside, outside, or notched, any shape you want.

Step 6. Pre-Surfacing Preparation

Challenge: to accomplish all tasks that need to precede surfacing the walls, as dictated by the surfacing material chosen, decisions about use of moisture barriers, reinforcement, etc.

The techniques are the same as used in preparing loadbearing walls for surfacing, except that, in some cases, you will have vertical "posts" to attach wire reinforcement to.

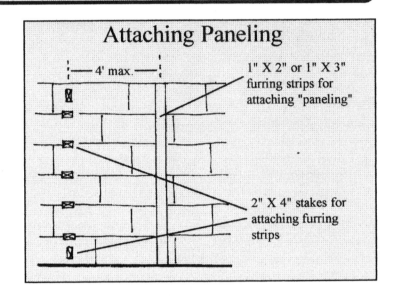

Attaching Paneling

— 4' max. —

1" X 2" or 1" X 3" furring strips for attaching "paneling"

2" X 4" stakes for attaching furring strips

Step 7. Surfacing the Bales

Challenge: same challenge, same considerations, same options as with loadbearing walls.

Step 8. Finishing Touches

Challenge: same as with loadbearing, including resisting the temptation to move in until the interior is really completed. Once you're moved in, life has a way of providing what seem to be higher priorities than getting that mortician-gray concrete floor stained and waxed, or caulking and painting in the bedroom (nobody but you sees it anyway, right?). Reconfirm your wedding vows; your relationship has survived a formidable test!

The Hybrid Option

Think about combining loadbearing bale walls with loadbearing frameworks (or elements made from other rigid materials) in your design. Tempting, eh?

Combinations of the two approaches enable you to escape, at least partially, some of the constraints of each. It requires attention to detail, however, since we will be mixing walls that compress under load with walls or elements that do not. Draw it, model it, think it through; be prepared to start building only to get blind-sided by some problem that escaped even your keen mind. Be ready to think on your feet. Take it to the limit.

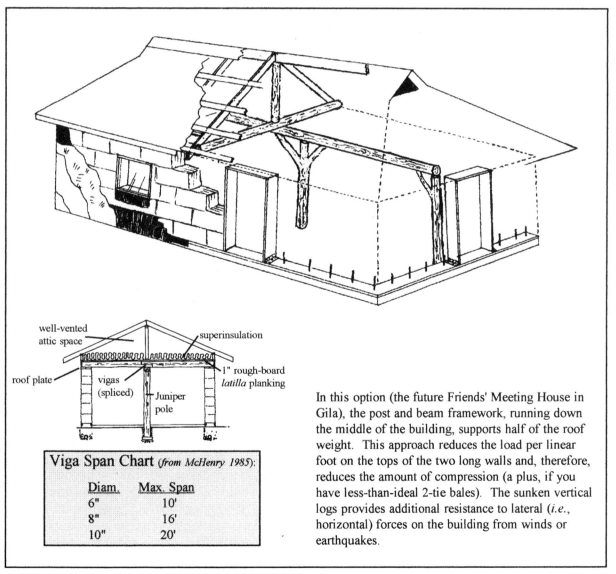

well-vented attic space

superinsulation

roof plate

vigas (spliced)

Juniper pole

1" rough-board *latilla* planking

Viga Span Chart *(from McHenry 1985)*:	
Diam.	Max. Span
6"	10'
8"	16'
10"	20'

In this option (the future Friends' Meeting House in Gila), the post and beam framework, running down the middle of the building, supports half of the roof weight. This approach reduces the load per linear foot on the tops of the two long walls and, therefore, reduces the amount of compression (a plus, if you have less-than-ideal 2-tie bales). The sunken vertical logs provides additional resistance to lateral (*i.e.*, horizontal) forces on the building from winds or earthquakes.

Structural Combinations

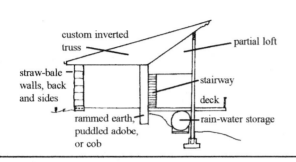

custom inverted truss

partial loft

straw-bale walls, back and sides

stairway

deck

rammed earth, puddled adobe, or cob

rain-water storage

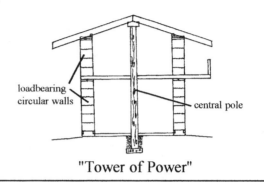

loadbearing circular walls

central pole

"Tower of Power"

Mixed Materials

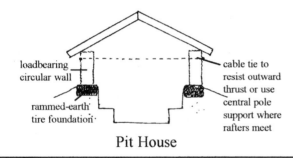

loadbearing circular wall

cable tie to resist outward thrust or use central pole support where rafters meet

rammed-earth tire foundation

Pit House

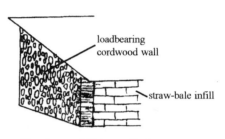

loadbearing cordwood wall

straw-bale infill

Cordwood/Straw Bale

Retrofits

Commercial Metal or Pole-building Kit

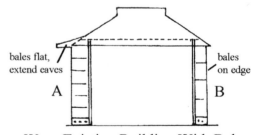

bales flat, extend eaves

bales on edge

A

B

Wrap Existing Building With Bales

Pay-As-You-Go Hybrid

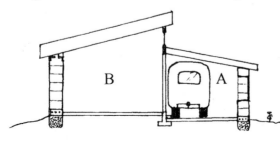

B

A

Phase 0 - Live in trailer, saving money to build a hybrid straw-bale home.
Phase 1 - Build space A to shade trailer, leaving one end temporarily plugged with bales.
Phase 2 - Save more money.
Phase 3 - Build Space B.
Phase 4. Unstack temporary end wall, remove trailer, replace end wall and finish.
Phase 5. Sell the trailer to someone who wants to do what you just finished doing.

Appendices

Photo by Matts Myhrman

Out On Bale workshop participants raising eight-course-high structural walls, using three-string bales layed flat.

Photo by Matts Myhrman

Loadbearing structure, with low-pitch gabled roof design, under construction in Southeastern Arizona.

Bamboo Pins

Eastern Star Trading Co.
624 Davis St.
Evanston, IL 60201
1-800-522-0085
Minimum bundle size – 500

Hummert's Seed Co.
2746 Chauteau St.
St. Louis, MO 63103
1-800-325-3055
Minimum bundle size – 200

Ceiling Insulation Alternatives

Air Krete
Palmer Industries, Inc.
10611 Old Annapolis Rd
Frederick, MD 21701

Cotton Insulation
Kirk Villar
Greenwood Cotton Insulation Products, Inc.
70 Mansell Court, Suite 100
Roswell, GA 30076
(404) 998-6888

Construction Tools

Bale Needles
Make your own or order from:
OOB-By Mail
1037 E. Linden St.
Tucson, AZ 85719

Hay Knives
Search them out in junk/antique stores.

Notch and Niche Cutter Blade – The Lancelot
King Arthur's Tools
3225 Earl Dr.
Tallahassee, FL 32308
1-800-942-1300

Water Level Kits
Widely available at hardware and building supply outlets. You supply the hose.

Foam-Core Structural Stress-Skin Panels

AFM Corporation
Box 246
Excelsior, MN 55331
(612) 474-0809
Known as R-Control Panels

"Green" and Healthy Building Materials

See CRBC (1994), St. John (1992), Harris (1994), Pearson (1989), and Dadd (1993)

Modeling Supplies

Askren Floral Supply
3501 E. Golf Links Road
Tucson, AZ 85713
(520) 325-7144
Straw-coated two-wire, styrofoam micro-bales (1" by 1.25" by 2.63")

Rhyne Floral Supply Mfg. Co.
P.O. Box 310
Gastonia, NC 28053
(704) 922-7841
As above, but sell only to businesses.

Moisture Measurement Devices

Protimeter, Inc.
P.O. Box 450
Danvers, MA 01923
1-800-321-4878
"Balemaster protimeter", calibrated to straw, hand-held with stanless-steel probe; 10%-70% range.

Delmhorst Instrument Co.
51 Indian Lane East
Towaco, NJ 07082
1-800-222-0638
*Hay moisture tester, 10" probe, digital readout,
13%-40% range is standard for the digital readout
meters but an alalog model is available for the
6%-30% range.*

Twine and Strapping

Carlson Systems
8990 F. St.
Omaha, NE 68127-1491
1-800-233-7447
Polyester strapping cord and wire buckles

Frank W. Winne & Son, Inc.
44 North Front St.
Philadelphia, PA 19106
(215) 627-6555
*240 lb. knot strength poly baling twine, Loktite
brand, and polyester strapping*

Straw Bales:
Regional Suppliers and Shippers

Carter's Tack and Feed
1700 W. Jagged Rock Rd.
Tucson, AZ 85737
(602) 887-6232

Albert Francis
7750 S. County Rd. 100
Alamosa, CO 81101
(719) 852-4642

Rick Green
2130 County Rd. "S"
Willows, CA 95988
(916) 934-7225
mobile (916) 952-8142

Navajo Agricultural Products Industries
Buddy Benally, Marketing Director
(505) 327-5251

Sioux Hay
Box 8
Meckling, SD 57044
Lawrence Opdahl at 1-800-303-3215

Mike Skinner
Box 982
Spearman, TX 79081
1-800-EASYHAY
*Two- and three-tie bales, including organically
grown wheat.*

Pressed Strawboard

Stramit—USA
P.O. Box 885
Perryton, TX 79070
(806) 435-9303
FAX 806-435-4311
*Pressed strawboard used primarily for interior
partition walls.*

Tarps

Northern
P.O. Box 1499
Burnsvill, MN 55337-0499
Durable, silver-surface, many sizes

Toilets

Commercial
Sun-Mar Corp.
900 Hertel Ave.
Buffalo, NY 14216
(416) 332-1314
Composting types

Homemade
*See Reynolds (1994) to make a solar dessication type
See Van der Ryn (1978) to make a composting type*

Wire

*For through-bale ties and "Robert" pin wire, try
salvage yards and fencing material suppliers.*

Appendix
Human Resources*
Two

* NOTE: *This evolving list of Human Resources was compiled and is kept current by Out On Bale (1037 E. Linden St., Tucson, AZ 85719). No endorsement of named entities is intended nor is criticism implied of entities not listed.*

UNITED STATES

Alabama

Marley Porter. Design Troupe, 27 Crabapple Ct Still Waters, Dadeville, AL 36853. *Architects.*

Alaska

Monte Lamer, P.O. Box 277, Hill Top Rd., Healy, AK 99743. *General resource.*

Arizona

Robert Andrews, 1119 East Mitchell, Tucson, AZ 85719. *House plans.*

Perry Becker, 7777 East Main #159, Scottsdale, AZ 85251. *Design, permits, building assistance.*

Eric Black. Black and Associates, 1623 A, N San Francisco St., Flagstaff, AZ 86001. *Mortgage lending consultant/broker.*

DeHavillan Workshops, 1039 E Linden St., Tucson, AZ 85719. *Workshops nation-wide. Has a pool of architects, builders, and educators.*

Dan Dorsey, P.O. Box 41944, Tucson, AZ 85717. *Straw-bale design, workshops on code compliance.*

David Eisenberg. Development Center for Appropriate Technology (DCAT), 2702 E. Seneca, Tucson, AZ 85716. *Consulting, education, testing & research, networker.*

Carol Escott and Steve Kemble. Sustainable Systems Support, P.O. Box 318, Bisbee, AZ 85603. *Consultation, design, workshops, and informational materials including videos and booklets.*

Tim Farrant. Close Enuf Construction, P.O. Box 41991, Tucson, AZ 85717. *Privacy walls; consulting, design, estimates, building; workshops.*

Bill Ford, 1227 N. 3rd Ave., Tucson, AZ 85705. *Architect, design, planning, code compliance.*

John George, 2403 E. Drachman, Tucson, AZ 85719. *Experienced wall captain, installs windows, doors and roof systems, makes steel angle-iron lintels.*

Tom Greenwood, 19 E. 15th St., Tucson, AZ 85701. *Architect; stock plans.*

Thomas Hahn, Architect. The Sol Source, 2501 N Edgemere St., Phoenix, AZ 85006-1115. *Design/build; passive solar.*

Mark Hawes and Ralph Towl. Sunbale, 4122 E. Kilmer, Tucson, AZ 85711. *Design/build; active and passive solar systems.*

Bob Lanning, 19 E. 15th St., Tucson, AZ 85701. *Architect,* *design/build, testing and research, consultation, education, house plans.*

Don Larry. Design Troupe, 86 West University Suite 107-A, Mesa, AZ 85201. *Architect.*

Brian Lockhart, P.O. Drawer X, Bisbee, AZ 85603. *Architect, code compliance, general resource.*

Jill Lorenzini, 427 S. 4th Ave. #2, Tucson, AZ 85701. *Information, workshops, networker, especially with women.*

Out On Bale—By Mail, 1037 E. Linden St., Tucson, AZ 85719. *A general resource, education and information center; written material and videos.*

Jon Ruez, 2327 E. 1st St., Tucson, AZ 85719. *Builder/designer, consultation, workshops.*

Liss Spencer, 427 South 4th Ave. #2, Tucson, AZ 85701. *Construction, interior finish work.*

Athena and Bill Steen. The Canelo Project, HC1 Box 324, Elgin, AZ 85611. *Authors; non-profit organization, exploring living, building, and food production practices, including straw-bale construction.*

Paul Weiner, 19 E. 15th St., Tucson, AZ 85701. *Design/build consultants. Licensed contractor.*

Women Build Houses, 1050 S. Verdugo, Tucson, AZ 85745. *A network of women, referrals to women builders, architects, and tradespeople. Small tool library and information.*

California

Lynn Bayless. EOS Institute, 580 Broadway, Suite 200, Laguna Beach, CA 92651. *Education and resource center. Regional resources, referrals.*

Gary Boudreaux and LaJan McAlister, 184 Grove St., Nevada City, CA 95959. *Builders-coordinators.*

Polly Cooper/Ken Haggard. San Luis Sustainability Group, 617 Oakridge, San Luis Obispo, CA 93405. *Architecture, planning, passive solar, construction, research, consulting.*

Jan Fillinger. Savidge, Warren, and Fillinger, Architects, 1250 Addison Studio 102, Berkeley, CA 94702. *Architects, solar.*

Rick Green, 2130 County Rd S, Willows, CA 95988. *Builder, provider of rice-straw bales.*

Stanley Scholl, 161 McKnight Drive, Laguna Beach, CA 92651. *Civil/structural engineer. Former building official.*

John Swearingen. Skillful Means Builders, P.O. Box 207, Junction City, CA 96048. *General contractors, design/build.*

Bob Theis. Daniel Smith & Associates, 1107 Virginia St., Berkeley, CA 94702. *Architect, design for earthquake resistance, improved roof systems, California code.*

Stan Welch. SRW Construction, 1075 Montgomery Rd, Sebastopol, CA 95472. *Construction, seminars.*

Colorado

Jim Chiaro. Construction Alternatives, 755 4th St., Penrose, CO 81240. *Builder/designer, passive solar, seminars, workshops, assistance.*

Keith Conway, P.O. Box 336, Crestone, CO 81131. *Builder, workshops, consulting, design.*

Steve Gellatly, Dawn McGrath. Strawlorodo, 15811 C.R. 31, Mancos, CO 81328. *Build, design, equipment, supplies, workshops.*

Bruce Glenn. Bruce Glenn Design & Construction, P.O. Box 1264, Bayfield, CO 81122. *Builder, stock plans.*

Tom Luecke. StrawCrafters, 3785 Moorhead Ave., Boulder, CO 80303. *Design-build, consulting, workshops, plastering, polyester strapping and related tools; testing, Colorado code.*

Sean Thompson. Alternative Design Network, 50 Sunset Unit #1, Basalt, CO 81621. *Construction, energy and water systems.*

James P. Verkaik, AIA. Verkaik & Associates, Inc. 831 Royal Gorge Blvd., Suite 418, Canon City, CO 81212. *Architect/engineer design, code, banks, insurance.*

Ken Williams. StrawBuild, 389 Sopris Ave., Carbondale, CO 81623. *Information, hands-on assistance.*

Iowa

George Swanson. CADD America, Box 662, Fairfield, IA 52556. *Design/build; breathing wall systems, Faswall.*

Maine

Lisa Hawkins, RFD 3 Box 536-A, Belfast, ME 04915. *General resource, education.*

Maria Gail, P.O. Box 1044, Belfast, ME 04915. *Owner-builder, hands-on assistance.*

Massachussetts

Paul Lacinski, 801 N Poland, Conway, MA 01341. *General resource.*

Rosie Heidkamp, Shirley White, 135 Farley, Wendell, MA 01379. *General resource, carpentry, furniture building.*

Michigan

Scott Barkdoll. Skywoods Carpentry, 12750 Oakley, Honor, MI 49640. *Owner-builder assistance, carpentry, mill work, roofing, skylights.*

Wayne Appleyard. Sunstructures Architects, 201 E Liberty, Ann Arbor, MI 48104. *Architect, passive solar.*

Montana

Marilyn Cochran, 157 Blodgett Ln., Arlee, MT 59821. *Carpentry.*

Steve Loken. Center for Resouceful Building Technology, P.O. Box 3413, Missoula, MT 59806. *General resource.*

Ramie Pederson, P.O. Box 200, Emigrant, MT 59027. *Construction and coordination; written materials and videos.*

Jim Petersen. ResourcefulNest, P.O. Box 641, Livingston, MT 59047. *Build/design.*

Alan Roy or Jamie Pennington, 1000 N 17th #35, Bozeman, MT 59715. *General resource.*

Nevada

Constance Alexander, Christy Tews, P.O. Box 364; Minden, NV 89423. *Information, workshops, consultation, plans.*

New Hampshire

Joe Battle. Battle Farms, Box 328 Center Rd., Bradford, NH 03221. *Source of straw bales.*

Michael Champagne, 13 FoxMeadow, Nashua, NH 03060. *General resource.*

New Mexico

Tony Beleen, P.O. Drawer 1318, Farmington, NM 87499. *General contractor; works for Navajo Agricultural Products Industry, a source of straw bales.*

Sumara Buchanan. Straw Bale Walls and Property Entries, Rt. 2, Box 422, Santa Fe, NM 87505. *Privacy walls and entrances. General resource.*

Danny Buck. Living Structures, Inc., P.O. Box 6447, Santa Fe, NM 87502. *Design, development, construction, landscaping.*

Virginia Carabelli, P.O. Box 660, Tesuque, NM 87574. *General consulting.*

Robert Collins. Strawbale Design, 1676 Cerro Gordo Rd., Santa Fe, NM 87501. *Steel frames, general resource.*

Meg and Randy Milligan, 3324 Pike Ct., Carlsbad, NM 88220. *General resources.*

Molly Nieman, P.O. Box 3572, El Prado, NM 87529. *Architect, design, consulting, education.*

Tony Perry. Straw-Bale Construction Management, Inc., 31 Old Arroyo Chamiso, Santa Fe, NM 87505. *Consulting, affordable housing, workshops, seminars.*

Betsy C. Pierce. Alternative Building Associates, 95 Moya Rd., Santa Fe, NM 87505. *Architect, contractor.*

MacLaren Scott and Paul Zelizer, HC Box 9503, Ranchos de Taos, NM 87557. *Design consulting, coordination, management.*

Beverley Spears, AIA. Spears Architects, 1334 Pacheco St., Santa Fe, NM 87501. *Architects/design.*

Straw Bale Construction Association, 31 Old Arroyo Chamiso, Santa Fe, NM 87505. *Association of architects, designers, engineers, contractors, subcontractors; code; sharing technical information.*

Ted Varney, 2308 Calle Luminoso, Santa Fe, NM 87505. *Builder.*

Cadmon Whitty, 902 6th NW, Albuquerque, NM 87102. *Privacy walls, buildings, workshops, general resource.*

New York

Hilary Oak and Chris Affre. Building With Bales In The Adirondacks, 2682 White Hill Rd., Parishville, NY 13672. *Information, networking.*

North Carolina

Susan Clellen, 416 Summit St., Walnut Cove, NC 27052. *Carpenter/builder.*

Mitchel Sorin, Rt. 1, Box 61-l, Whittier, NC 28789. *Architect, design, code compliance, technical assistance, alternative building materials.*

Ohio

Ken Matesz. Fountainhead Natural Homes, 11965 Monclova Rd., Swanton, OH 43558. *Construction, consulting, workshops.*

Oregon

Linda Barnes, 1231 NW Hoyt #403, Portland, OR 97209. *Architect, code compliance, general resource.*

Justyn Livingston, P.O. Box 497, Camp Sherman, OR 97730. *General resource.*

Kyle MacLowry, 132 NE 55th, Portland, OR 97213. *General resource.*

Andre DeBar, 246 NE Thompson, Portland, OR 97212. *Architect, passive solar.*

Southern Oregon Straw (SOS), c/o ESRI, P.O. Box 627, Ashland, OR 97520. *General resource.*

South Carolina

Cal Collins. Amber Waves Construction, 1216 Oakland Ave., Florence, SC 29506. *General resource.*

Tennessee

Howard Switzer, 219 Bonnafield, Hermitage, TN 37076. *Architect, design/build/consult.*

Texas

Duncan Echelson. Bowerbird Construction, P.O. Box 698, Dripping Springs, TX 78620. *Build, consult.*

Pliny Fisk III. Center for Maximum Potential Systems, 8604 FM 969, Austin, TX 78724. *Building and design center, large projects, Texas code.*

Pamela Overeynder. Foundation for a Compassionate Society, 227 Congress Suite 370, Austin, TX 78701-4021. *General resource.*

Utah

Dave Clark and LeeAnn Truesdell, 708 Locust Lane, Moab, UT 84532. *General resource.*

Virginia

R. Teixeira and A. Adams. Bale-Out, RR 3 Box 77-A, Floyd, VA 24091. *Consulting, education and workshop techniques.*

Washington

Ted Butchart, Peggy Robinson. Greenfire Institute, 1509 Queen Anne Ave. N. #606, Seattle, WA 98109. *Build/design, workshops, general resource, permaculture.*

Gary Headlee, Rt. 2, Box 99-G, Omak, WA 98841. *Builder/contractor, code, general resource.*

Simon Henderson , Larry Santoyo. Great NW Permaculture, 2073 Marble Valley Rd., Addy, WA 99101. *Education, networking.*

Chris Stafford and B.J. Harris. Stafford-Harris, Inc., 1916 Pike Place #705, Seattle, WA 99101. *Architects, design, consulting.*

Wisconsin

Dreamtime Village, Rt. 2, Box 242W, Viola, WI 54664. *Design/build/teach, solar, general resource.*

Frederick Lehmann. Lehmann Construction, W 6272 72nd Ave., Beldenville, WI 54003. *General contractor.*

Wyoming

Chip Williams, 651 Middlefork Ln., Lander, WY 82520. *Carpentry, computer-aided drafting.*

CANADA

Alberta

Jorg and Helen Ostrowski. Alternative and Conservation Energies (ACE), 9211 Scurnfield Dr. NW, Calgary, Alberta T3L 1V9 Canada. *Design/build, planning, consulting, R&D.*

Nova Scotia

Shawna Henderson. Abri, 106 Silver Birch Drive, Armdale, Nova Scotia B3L 4J3 Canada. *Owner-builder, computer evaluation.*

Kim Thompson. Straw House Herbals, RR#1 Ship Harbour, Nova Scotia B0J 1Y0 Canada. Consultant/designer, workshops, Ship Harbour project video and book.

Ontario

Linda Chapman, 27 Third Ave., Ottowa, Ontario K1S 2J5 Canada. *Architect, permaculture.*

Krystopher J. Moreau. Eagle Dawn Design, 22 Suraty Ave., Scarborough, Ontario M1P 2E1 Canada. *Design, general resource, code compliance.*

Tom Ponessa, 412 Palmerston Blvd., Toronto, Ontario M6C 2N8 Canada. *Architect, builder, teacher.*

Quebec

Francois Tanguay, Michel Bergeron, Clode De Guise. ARCHIBIO, L'Eperviere, 1267 Chemen Lac Deligny Quest, Mandeville, Quebec J0K 1L0 Canada. *Design/build, workshops, consultations, author.*

ENGLAND

Barbara Jones. Amazon Nails-Women Roofers and Building Contractors, 554 Burnley Rd., Todmorden 0L148JF England. *Roofer, joiner, researcher, advocate.*

FINLAND

Tapani Marjamaa, Sinikontie 3, c/o Keijo Marjamaa, 74300 Sonkajarvi, Finland. *Built first straw bale in Finland; leading testing, research effort for national code.*

FRANCE

John Daglish. Biotique Habitat, 122 Ave. Saint Exupery Antony 92160 France. *Architect, building biologist, permaculture designer, general resource.*

Pascal Thepaut, Trovoas 29640, Plougonven, France. *Architect, bau-biologist.*

NEW ZEALAND

Peter Kundycki, 3b Telford Tce, Wellington 1, New Zealand. *Design/Build passive solar, permaculture.*

Code, Testing and Research

Three

Building Code

How the typical owner-builder feels about "the code" was neatly described by David Eisenberg in the Winter,1994 issue of *The Last Straw* journal:

"Among the dreaded things in life, right up there with death and taxes, is dealing with permits, plans, building codes and inspectors. When you're planning to build with straw bales, or any unusual or innovative material or method, the prospect of negotiating the maze of 'the codes' seems even more ominous. Don't give up hope."

The purpose of the Uniform Building Code (and, generally, of all such codes) is, in its own words, "to provide minimum standards to safeguard life or limb, health, property and public welfare by regulating and controlling the design, construction, quality of materials, use and occupancy, location and maintenance of all buildings and structures within this jurisdiction and certain equipment specifically regulated herein".

Primary focus by enforcement officials is usually focused on "life or limb, [and] health". Most of us would probably agree that the intention, as stated here, is legitimate, although we might prefer to be allowed to individually take responsibility for this aspect of buildings that we create for ourselves.

The straw-bale builder will find no specific details on how one can build safely with bales in the standard version of any of the several building codes used in the USA. However, there is a lifesaver to be found in Sections 105, 106 and 107 of the UBC. Section 105 states, in part, "The provisions of this code **are not intended** [emphasis added] to prevent the use of any material or method of construction not specifically prescribed by this code, provided any alternate has been approved and its use authorized by the building official."

Separate processes are nearing completion in New Mexico and in the cooperating jurisdictions of Pima County and the City of Tucson, Arizona, which would incorporate prescriptive standards for straw-bale construction into their respective versions of the UBC. In New Mexico the standard will allow only non-structural use of bales, whereas in Pima County and Tucson, loadbearing bale walls will also be permitted.

The progress made toward getting the technique formally prescribed in the code has depended largely on:
- the perseverance and credibility of Tony Perry and the Straw Bale Construction Association in New Mexico and of David Eisenberg and Matts Myhrman in Arizona.
- the genuine cooperation and involvement of enlightened building officials in all three jurisdictions, and

- the results of testing programs, described below, which were supported entirely by the donation of labor and cash by individual "*strawficionados*".

Testing

Tucson, Fall, 1992 – The structural testing of individual, unconfined bales and of panels of unstuccoed, three-tie bales laid flat is described in a University of Arizona master's thesis by Ghailene Bou-Ali (civil engineering). The panels were loaded vertically (compression), and horizontally (both parallel with the plane of the walls and at a right angle to them). The process and results were summarized in the Summer,1993 issue of *The Last Straw* by David Eisenberg. His more comprehensive summary, published by the Community Information Resource Center is available from Out On Bale–By Mail (see *Appendix Two*).

Tucson, Spring, 1993 – Using a modified hot plate heat source technique, University of Arizona graduate student in energy engineering, Joe McCabe, determined the R-value per inch for wheat straw in bales having a calculated density of just over 8 pounds per cubic foot. For heat moving parallel to the straws, the value was 2.38, while for heat moving in a direction perpendicular to the straws the value was 3.0. A more detailed summary of his results appeared in the Summer,1993, issue of *The Last Straw*.

New Mexico, Winter, 1993, Spring, 1994 – Structural testing relevant to non-loadbearing use of bales (*i.e.*, horizontal force at a right angle to the wall plane, ASTM E-330) and a small-scale fire test (ASTM E-119) were performed by SHB AGRA, Inc., a commercial testing laboratory in Albuquerque, New

Mexico. Additional testing for thermal conductivity (to permit determination of R-value per inch for baled straw) was conducted by Sandia National Labaoratories. A packet containing descriptions and the very positive results of these tests is available from the Straw Bale Construction Association and from OOB-By Mail (see *Appendix Two*). An abbreviated wish list for further testing includes:

- structural testing of wall panels of two-tie bales
- structural testing of wall panels (two- and three-tie bales) pinned with bamboo instead of rebar
- structural testing of three-tie bale wall panels with external, polyester strapping, roof plate tie-downs rather than internal, threaded rod tie-downs
- determination of the relationship of R-value per inch to degree of compaction for bales and for loose straw packed in bags for ceiling insulation
- full scale fire testing (ASTM E-119) of stuccoed and mud-plastered straw-bale walls.

Research and Development

An abbreviated wish list for research and development includes:

- development of an inexpensive, compact, lightweight, standardizable device for directly measuring the degree of compaction of baled materials, regardless of their moisture content
- determination of the conditions, if any, under which spontaneous combustion is possible in straw- or hay-bale walls
- determination of whether damaging, in-wall condensation does occur in straw-bale walls in houses located in climatic regions characterized by high rainfall and humidity

and by cold winter temperatures. If such condensation does occur and does support breakdown of the straw by fungi, development of strategies to prevent this from happening will be needed.

Your Help is Needed

Since straw-bale construction belongs to all of us (*i.e.*, no one of us), the initial financial support for testing, research and development must also come from all of us. As more credibility is earned and gained, we will be able to more effectively put pressure on the federal government to do its part.

You can support this initial testing, research and development by:

- Using your network to promote the need for such activities and to identify individuals or entities that might wish to help with these efforts.

- Sending a tax-deductible contribution to these funds:

  DCAT-BRAN
  Straw-bale Research and Testing Fund
  P. O. Box 41144
  Tucson, AZ 85717
  Specializing in small seed-grants, often to graduate students, for research and development of straw-bale-related technology, small-scale testing and documentation of the historic tradition and techniques.

  NMCF-AHEF-Straw Bale
  c/o Danny Buck, Treas.
  SBCA
  1334 Pacheco St.
  Santa Fe, NM 87505
  Focusing presently on further structural testing by commercial laboratories.

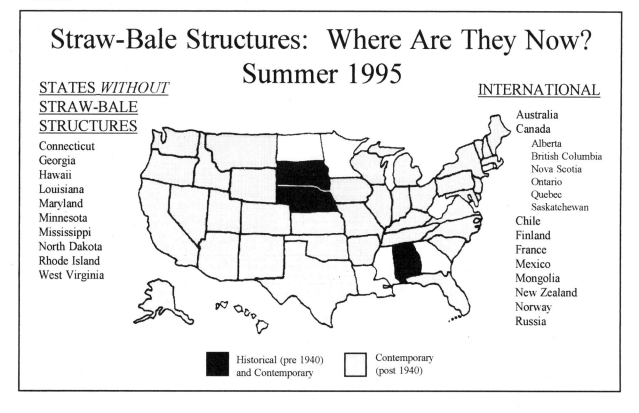

Straw-Bale Structures: Where Are They Now?
Summer 1995

STATES *WITHOUT* STRAW-BALE STRUCTURES

Connecticut
Georgia
Hawaii
Louisiana
Maryland
Minnesota
Mississippi
North Dakota
Rhode Island
West Virginia

INTERNATIONAL

Australia
Canada
   Alberta
   British Columbia
   Nova Scotia
   Ontario
   Quebec
   Saskatchewan
Chile
Finland
France
Mexico
Mongolia
New Zealand
Norway
Russia

■ Historical (pre 1940) and Contemporary

☐ Contemporary (post 1940)

"I OBTAINED A LIST OF NEW HOUSING STARTS AND NOW I'M MATCHING IT WITH LARGE PURCHASES OF STRAW BALES & BALING TWINE.....LOOK OUT PIGS, HERE WE COME!"

Post-and-beam, bale-insulated mansion near Huntsville, Alabama. Finished in 1938, Dr. Burritt's home is now maintained as a museum. The annual rainfall averages about 50 inches.

Reading and Viewing

Hybrid design with straw-bale walls carrying half the weight of each shed roof.

Virginia Carabelli's completed post-and-beam, bale-insulated home near Santa Fe, New Mexico. First residence ever built with permit. See also photo on page 5.

Suggested Reading and Viewing

Books

Bainbridge, D., with B. Steen and A. Steen. 1993. Plastered straw bale construction. The Canelo Project, Elgin, AZ. 51p.

Brower, S. N. 1988. Design in familiar places: what makes home environments look good. Praeger, NY. 187 p.

Ching, F. 1991. Building construction illustrated. Van Nostrand Reinhold, NY. Unpaginated.

Creasy, R. 1982. The complete book of edible landscaping. Sierra Club Books, San Francisco, CA. 379 p.

Dietz, A. G. H. 1991. Dwelling house construction. Fifth edition. Cambridge, Mass. 440 p.

Dickinson, D. 1986. The small house: an artful guide to affordable residential design. McGraw-Hill, NY. 196 p.

Hop, F. 1989. The energy-saving house design handbook: your super-guide to earth sheltering, solar heating and thermal construction. Prentice Hall, NY. 216 p.

Kidder, T. 1985. House. Houghton Mifflin, Boston. 341 p.

Kilpatrick, J. 1989. Understanding house construction. Home Builder Press of NAHB, Wash. D.C. 79 p.

Lanning, B. 1995. Straw bale portfolio: a collection of sixteen designs for straw bale houses. Self-published. Available from OOB-By Mail (*see Appendix Two*).

Levy, M. and M. Salvadori. 1992. Why buildings fall down: how structures fail. W. W. Norton, NY. 334 p.

Mazria, E. 1979. The passive solar book: a complete guide to passive home, greenhouse, and building design. Rodale Press, Emmaus, PA. 435 p.

Merritt, F. S. and J. T. Ricketts. 1994. Building design and construction handbook. McGraw-Hill, NY.

Mollison, B. C. 1991. Introduction to permaculture. Tagari Publ., Tyalgum, Australia. 198 p.

Reynolds, M. 1991. Earthship, volume II: systems and components. Solar Survival Press, Taos, NM. 255 p.

Sardinsky, R. and the Rocky Mountain Institute. 1992. The efficient house sourcebook. The Rocky Mountain Inst., Old Snowmass, CO. 165 p.

Steen, A., B. Steen, D. Bainbridge, with D. Eisenberg. 1994. The straw bale house. Chelsea Green Publ. Co., White River Jct. 297 p.

Warren, J. 1984. Building a fortune: the book that new house. Garnet Publ. Co., Spokane, WA. 164 p.

Webster, D. 1987. How to save money when you build your house: a guide for people who want to build their own house on a limited budget. Triangle Publ., Rose Hill, KS. 132 p.

Wylde, M., A. Baron-Robbins, and S. Clark. 1993. Building for a lifetime: the design and construction of fully accessible homes. The Taunton Press, Newtown, CT. 295 p.

Magazines & Journals

Environmental Building News. EBN, RR 1, Box 161, Brattleboro, VT 05301. (802) 257-7300. *Bi-monthly.*

Fine Homebuilding. The Taunton Press, Box 9974, Newtown, CT 06470-9974. *Bi-monthly.*

Home Power. HP, P.O. Box 520, Ashland, OR 97520. (916) 475-0830. *Bi-monthly.*

The Journal of Light Construction. JLC, P.O. Box 686, Homes, PA 19043. 1-800-345-8112. *Monthly.*

The Last Straw. Out On Bale, 1037 E. Linden St., Tucson, AZ 85719. *Quarterly.*

Videos

Building With Straw: Volume 1–A Straw-bale Workshop; Volume 2–A Straw-bale House Tour. Black Range Films, Star Route 2, Box 119, Kingston, NM 88042.

Straw Bales for Shelter. Out On Bale, 1037 E. Linden St., Tucson, AZ 85719.

How to Build Your Elegant Home with Straw Bales. Sustainable Support Systems, Box 318, Bisbee, AZ 85603.

[*All of the above available from* OOB-By Mail–*see Appendix Two.*]

Literature Cited

Adelman, D. 1984. Radiant-floor heating. Fine Homebuilding, 8/84:68-71.

Alexander, C. 1977. A pattern language. Oxford University Press, New York. 1171p.

_____. 1979. The timeless way of building. Oxford University Press, New York. 352p.

Alfano, S. 1985. The art and science of estimating. Fine Homebuilding, 6/85:32-35.

Anderson, B., and M. Wells. 1994. Passive solar energy: the homeowner's guide to natural heating and cooling. 2nd edn. Brick House Publ. Co., Amherst, NH. 168p.

Ballard, S. 1987. How to be your own architect: a residential design handbook. Betterway Publications, White Hall, VA. 235p.

Benson, T. 1988. The timber-frame home: design, construction and finishing. The Taunton Press, Newtown, CT. 240p.

Berglund, M. 1985. Soil-cement tile floor. Fine Homebilding, 6/85:56-59.

Booth, D., J. Booth, and P. Boyles.1983. Building for energy independence: sun-earth buffering and super-insulation. Community Builders, Canterbury, NH. 219p.

Bower, J. 1993. Healthy house building: a design and construction guide. The Healthy House Institute, Unionville, IN. 384p.

Breecher, M. 1992. Healthy homes in a toxic world: preventing, identifying and eliminating hidden health hazards in your home. Wiley, New York. 246p.

Cecchettini, P., J. Wood, and B. Beckstrom. 1989. Home plans for solar living. Home Planners, Inc., Farmington Hills, MI. 192p.

Clegg, P. and D. Watkins. 1987. Sunspaces. Storey Communications, Pownal, VT. 206p.

Connell, J. 1993. Homing instinct: using your lifestyle to design and build your home. Warner Books, New York. 404p.

Cook, J. 1989. Passive cooling. MIT Press, Cambridge, MA. 593p.

CRBC (Center for Resourceful Building Technology). 1994. Guide to resource efficient building elements. Center for Resourceful Building Technology, Missoula, MT. 94 p.

Curran, J. 1979. Drawing plans for your own home. Brooks Publ. Co., Bakersfield, CA. 227p.

Dadd, D. L. 1990. Non-toxic natural and earthwise: how to protect yourself and your family from harmful products and live in harmony with the earth. Jeremy P. Tarcher, Los Angeles. 360 p.

Davidson, J. 1990. The new solar electric home. AATEC Publs., Ann Arbor, MI. 408p.

Day, C. 1990. Places of the soul: architecture and environmental design as a healing art. Aquarian Press, Wellingborough, England. 192 p.

Doolittle, B. 1973. A round house of straw bales. Mother Earth News 19:52-57.

Dunkley, D. 1982. Putting the roof on. Fine Homebuilding, 8/82:64-69.

Eisenberg, D. 1993. Results of a structural straw bale testing program. Community Information Resource Center, Tucson, AZ. Distributed by OOB–By Mail (see *Appendix Two*). 11 p.

_____. 1995. Straw-bale construction and the building codes: a working paper. Development Center for Appropriate Technology, Tucson, AZ. 30 p.

Erickson, J. 1989. The homeowner's guide to drainage control and retaining walls. Tab Books, Blue Ridge Summit, PA. 152p.

Erley, D., and M. Jaffe. 1979. Site planning for solar access: a guidebook for residential developers and site planners. U.S. Dept. of Housing and Urban Development (DOE). 149p.

Feirer, J., and G. Hutchings. 1986. Carpentry and house construction. Bennett and McKnight, Encino, CA. 1120p.

Feirer, M. 1986. Making a structural model. Pgs. 2-5 in Construction techniques 2. The Taunton Press, Newtown, CT. 232p.

Gibson, S. 1994. Air and vapor barriers. Fine Homebuilding, 4/94:48-53.

Gorman, J. R., S. Jaffe, W. Pruter, and J. Rose. 1988. Plaster and drywall systems manual. 3rd edn. BNI Books , Los Angeles, CA, distributed by McGraw-Hill, New York. 415p.

Groesbeck, W., and J. Striefel. 1994. The resource guide to sustainable landscapes. Environmental Resources, Inc.. Salt Lake City, UT. 152 p.

Gross, M. 1984. Roof framing. Craftsman Book Co., Carlsbad, CA 475 p.

Hageman, J. 1991. Contractor's guide to the building code. Craftsman Book Co. 544 p.

Harris, B. J. Ongoing. The Harris directory: recycled content building materials. Disk for PC's. The Stafford Architects, 1916 Pike Place #705, Seattle, WA 98101-1056 (tel 206/682-4042).

Herbert, R. D. 1989. Roofing: design criteria, options, selection. R. S. Means, Kingston, MA 223 p.

HUD (US Dept. of Housing and Urban Development). 1995. Design guide for frost-protected shallow foundations. Available from HUD-User, P. O. Box 6091, Rockville, MD 20849. Call toll-free at 1-800-245-2691.

Hughes, K. 1987. Gunite retaining walls. Fine Homebuilding, 6/87:60-63.

Jackson, M. 1990. The good house: contrast as a design tool. The Taunton Press, Newtown, CT. 147 p.

Jackson, W. P. 1979. Building Layout. Craftsman Book Co., Solana Beach, CA. 238 p.

Jones, R. W., and R. D. McFarland. 1984. The sunspace primer: a guide for passive solar heating. Van Nostrand Reinhold, NY. 285 p.

Kern, K. 1975. The owner built home. Charles Scribner's Sons, New York. 374 p.

Kern, B., and K. Kern. 1981. The owner-built pole frame house. Scribner, NY. 179 p.

Law, T. 1982a. Roof framing simplified. Fine Homebuilding, 8/82:62-63.

_____. 1982b. Site layout. Fine Homebuilding, 10/82:26-28.

Lechner, N. 1991. Heating, cooling, lighting: design methods for architects. Wiley, NY. 524 p.

Lenchek, T., C. Mattock, and J. Raabe. 1987. Superinsulated design and construction: a guide for building energy-efficient homes. Van Nostrand Reinhold, NY. 172 p.

Levin, A. H. 1991. Hillside building: design and construction. Arts & Architecture Press, Santa Monica, CA. 172 p.

Loy, T. 1983. Understanding concrete. Fine Homebuilding, 2/83:28-32.

Lstiburek, J., and J. Carmody. 1993. Moisture control handbook: principles and practices for residential and small commercial buildings. Van Nostrand Reinhold, New York. 214p.

Luttrell, M. 1985. Warm floors. Fine Homebuilding, 6/85:68-71.

Lynch, K., and G. Hack. 1984. Site planning. MIT Press, Cambridge, Mass. 499 p.

McElderry, W., and C. McElderry. 1979. Happiness is a hay house. Mother Earth News 58:40-43.

McHarg, I. L. 1969. Design with nature. For Amer. Mus. Nat. Hist. by Natural History Press, Garden City, NY. 197 p.

McHenry, P. G. 1985. Adobe: build it yourself. Univ. of Arizona Press, Tucson. 158 p.

_____. 1989. Adobe and rammed earth buildings: design and construction. Univ. Arizona Press, Tucson. 217 p.

Moffat, A., Schller, and the staff of Green Living. 1993. Energy-efficient and environmental landscaping. Appropriate Solutions Press, South Newfane, VT. 230 p.

Mollison, B. C., and R. M. Slay. 1988. Permaculture: a designers' manual. Tagari, Tyalgum, Australia.

Monahan, E. J. 1986. Construction of and on compacted fills. Wiley, NY. 200 p.

MWPS (Midwest Plan Service). 1989a. Designs for glued trusses. Midwest Plan Service, Ames, IA. 84 p.

_____. 1989b. Farm and home concrete handbook. MPWS-35. Iowa State University, Ames, IA. 46p.

Nabokov, P., and R. Easton. 1989. Native American architecture. Oxford University Press, New York. 431 p.

Nisson, J. D. N., and G. Dutt. 1985. The superinsulated home book. Wiley, NY. 306 p.

NRAES (Northeast Regional Agricultural Engineering Service). 1984. Pole and post buildings: design and construction handbook. NRAES-1. Available through Midwest Plan Service, Iowa State University, Ames, Iowa. 47 p.

Pacey, A., and A. Cullis. 1986. Rainwater harvesting: the collection of rainfall and run-off in rural areas. Intermediate Technology Publ., London. 216 p.

Pearson, D. 1989. The natural house book: creating a healthy, harmonius & ecologically sound home environment. Simon and Schuster, NY.

Pegg, B. F., and W. D. Stagg. 1976. Plastering: a craftsman's encyclopaedia. Crosby Lockwood Staples, London. 276 p.

Potts, M. 1993. The independent home: living well with power from the sun, wind and water. Chelsea Green Publ. Co., Post Mills, VT. 300 p.

Reynolds, M. 1990. Earthship, volume I: how to build your own. Solar Survival Press. Taos, New Mexico. 229 p.

_____. 1994. Earthship III: evolution beyond economics. Solar Survival Press, Taos, NM.

Ring, D. 1990. Brick [on sand] floors. Pp. 94-97, in Foundations and masonry (J. Lively, ed.). The Taunton Press, Newtown, CT. 127 p.

Rybczynski, W. 1989. The most beautiful house in the world. Viking, New York. 211 p.

Strang, G. 1983. Straw bale studio. Fine Homebuilding, 12/83:70-72.

Steen, A. S., B. Steen, D. Bainbridge, with D. Eisenberg. 1994. The straw bale house. Chelsea Green Publ. Co., White River Junction, VT. 297 pp.

SBCA (Straw Bale Construction Association). 1994. The New Mexico engineering tests, thermal conductivity testing and draft building code. Straw Bale Construction Association, Santa Fe, NM. Various pagings.

Sherwood, E., and R. Stroh. 1992. Wood frame house construction: a do-it-yourself guide. Sterling Publ. Co., NY. 306 p.

Smulski, S. 1994. All about roof trusses. Fine Homebuilding, 6/94:40-45.

Southwick, M. 1981. Build with adobe. 2nd edition. Sage Books, Swallow Press, Athens, OH. 225 p.

Spence, W. P. 1993. Architectural working drawings: residential and commercial buildings. Wiley, NY. 521 p.

St. John, A. 1992. The sourcebook for sustainable design. Architects for Social Responsibility. ASR, Boston, MA.

Strong, S. 1994. The solar electric house book: a design manual for home-scale photovoltaic power systems. Sustainability Press, Still River, MA. 288 p.

Syvanen, B. 1982. Drafting: tips and tricks on drawing and designing house plans. East Woods Press, Charlotte, NC. 111 p.

_____. 1983. Small-job concrete. Fine Homebuilding, 2/83:34-35.

_____. 1986. Insulating and parging foundations. Pp. 179-181, *in* Construction techniques 2. The Taunton Press, Newtown, CN. 232 p.

Talcott. C., D. E. Helper, and P. R. Wallach. 1986. Home Planner's guide to residential design. McGraw-Hill, New York. 218 p.

Taylor, J. S. 1983. Commonsense architecture: a cross-cultural survey of practical design principles. W. W. Norton, NY. 160 p.

Thoreau, H.D. 1950. Walden - Civil disobedience. Rinehart & Co., New York. 304 p.

Tibbets, J. 1989. The earthbuilders' encyclopedia. Southwest Solaradobe School, Bosque, NM. 196 p.

Van der Ryn, S. 1978. The toilet papers: designs to recycle human waste and water. Capra Press, Santa Barbara, CA. 127 p.

Velonis, E. 1983. Rubble-trench foundations. Fine Homebuilding, 12/83:29-31..

Wahlfeldt, B. 1988. Wood frame house building, an illustrated guide. Tab Books, Blue Ridge Summit, PA. 262 p.

Weidhaas, E. R. 1989. Architectural drafting and construction. Allyn and Bacon, Boston. 575 p.

Welsch, R.L. 1970. Sandhill baled-hay construction. Keystone Folklore Quarterly, Spring Issue: 16-34.

_____. 1973. Baled hay. Page 70, *in* Shelter (L. Kahn, ed.). Shelter Publications, Bolinas, California. 176 p.

Wolfe, R. 1993. Low-cost pole building construction. Garden Way Publishing, Charlotte, Vermont. 178 p.

Wright, F. L. 1954. The natural house. Horizon Press, NY. 223p.